The Dead Beat

*Lost Souls, Lucky Stiffs, and
the Perverse Pleasures of Obituaries*

Marilyn Johnson

HarperCollins*Publishers*

FIRST EDITION

Designed by Joseph Rutt

Printed on acid-free paper

Library of Congress Cataloging-in-Publication Data

Johnson, Marilyn.
    The dead beat : lost souls, lucky stiffs, and the perverse pleasures of obituaries / Marilyn Johnson.—1st ed.
        p.   cm.
ISBN-10: 0-06-075875-9
ISBN-13: 978-0-06-075875-2
    1. Obituaries—History and criticism. I. Title.

PN4784.O22J65   2006
070.4'4992—dc22

2005052817

06 07 08 09 010   ❖/RRD   10 9 8 7 6 5 4 3 2 1

To Rob

Contents

The
Dead Beat

1

I Walk the Dead Beat

People have been slipping out of this world in occupational clusters, I've noticed, for years. Four journalists passed their deadline one day, and their obits filled a whole corner of the paper. What news sent them over the edge? How often do you see two great old actresses take their bows, or two major-league pitchers strike out together? Often enough to spook. Some days sculptors are called, some days pioneer cartoonists. A *New York Times* editor threw up his hands on June 13, 2004, and ran two almost perfectly parallel stories under one headline: WINNERS OF THE MEDAL OF HONOR FROM TWO ERAS DIE; BOTH MEN SAVED FELLOW MARINES.

It is more than coincidence, and certainly more than the vigilance of an editor on the graveyard shift. It's supernatural. I thrilled recently to a pair of obituaries for Paul Winchell, the voice of Tigger in *Pooh*, and John Fiedler, the voice of Piglet in *Pooh*; the two had gone silent a day apart. I keep them next to my clip from October 25, 1986, the day the *New York Times* ran side-by-side obituaries for the scientist who

isolated vitamin C and the scientist who isolated vitamin K.
One was ninety-three; the other ninety-two. One died on a
Wednesday, one on a Thursday. One's farewell ran three
columns, one ran two. One extracted the vitamin from tons of
cattle adrenals scooped from the Chicago slaughterhouses,
and also from paprika. One extracted female hormones from
tons of sow ovaries. Make something of these differences if
you dare. Albert Szent-Gyorgyi and Edward Adelbert Doisy,
Sr., Dr. C and Dr. K respectively, both Nobel Prize winners,
left the world together.

Did they get the idea from John Adams and Thomas Jefferson? In 1826, the second and third presidents of the United
States died in harmony on July 4, exactly fifty years after they
signed the Declaration of Independence. The *New-York
American* wrote:

> *By a coincidence marvellous and enviable, THOMAS JEF-
> FERSON in like manner with his great compeer, John
> Adams, breathed his last on the 4th of July. Emphatically may
> we say, with a Boston paper, had the horses and the chariot
> of fire descended to take up the patriarchs, it might have
> been more wonderful, but not more glorious. We remember
> nothing in the annals of man so striking, so beautiful, as the
> death of these two "time-honoured" patriots, on the jubilee of
> that freedom, which they devoted themselves and all that was
> dear to them, to proclaim and establish. It cannot all be chance.*

No, surely it cannot all be chance. These are mystical
forces, and what better place to find them at work than in the
obituaries?

Such coincidences don't occur every day, but it wouldn't take you a week to begin a creative collection. A veteran UPI photographer and a veteran AP photographer. A professor of theology, a pastor, and a nun. An author named Arthur, an architect named Aaron, and an artist named Alois. Two obstetricians. The inventor of alternate-side-of-the-street parking and one of the founders of Evelyn Wood's course in alternate-word reading. The service industry of Hollywood—a hairdresser, a caterer, and a costume designer. Princess Diana and Mother Teresa! Cary Grant and Desi Arnaz. The head of the tiniest kingdom in the world, the Vatican (Pope John Paul II), and the leader of the second-tiniest kingdom in the world, Monaco (Prince Rainier).

This is not craziness. It's careful newspaper reading. Each day, after I read, I wash the newsprint off my hands and think about universal harmonies. I think about things I haven't thought about since childhood, such as guardian angels. I used to believe we each walked around with a sort of ghost of ourself guiding and watching over us. Is it possible that instead of a guardian angel we each have a double, a guarantee that our work gets done? If we're the sort who isolates alphabet vitamins, there are two of us, just in case. If we are the voice of Tigger, the voice of Piglet backs us up.

A friend of mine used to collect "bus plunge" headlines. You'd be amazed how easy these are to collect. Buses plunge over cliffs and into canyons across the world, and newspaper editors seem resigned to the sameness and predictability of such a universal death. Nearly every headline reads, SO MANY KILLED IN SUCH AND SUCH COUNTRY'S BUS PLUNGE. Once, the *New York Times* reported 10 DIE IN BRAZIL BUS PLUNGE, though it wasn't even a bus that plunged. It was a truck. But the convention persists.

I think of bus plunges as the generic passing. Many of us took the plunge yesterday. What did we have in common? We happened to be riding the same bus. Perhaps the bus is literal—ten of us over a precipice in a south Brazilian state. Or perhaps it is metaphoric—an imaginary bus that on Saturday encapsulates two vitamin scientists and on Sunday bears a cargo of handmaidens to Hollywood.

The bus is an attempt to grasp the unthinkable, of course: one day we're riding along on the highway; the next, we plunge out of sight. Who knows who might be sitting beside us? Watergate prosecutor Archibald Cox's seatmate was Watergate counsel Sam Dash. Lawrence Welk's trumpeter and his accordion player played a duet out the door. The queen of the Netherlands and the king of the frozen french fry left the party together. The editor-in-chief of the *Bulletin of Atomic Scientists* went off with the lead guitarist for a rock group called the Blasters. I clipped them all.

The *New York Times* comes each morning in a blue plastic wrapper, and never fails to deliver news of the important dead. Every day is new; every day is fraught with significance. I arrange my cup of tea, prop up my slippers. I open the not-yet-smudged pages of newsprint. Obituaries are history as it is happening. I know one of these days, maybe even as these pages zoom off to the press, Nelson Mandela and Doris Day and Keith Richards will materialize in the obituaries. The fact that they haven't yet fills me with awe. I was sound asleep when

Picasso died ('73), absent for the death of Miles Davis ('91). I missed the end of their stories. I wasn't yet alive in my time.

Whose time am I living in? Was he a success or a failure, lucky or doomed, older than I am or younger? Did she know how to live? I shake out the pages. Tell me the secret of a good life!

Other people, it seems, also read the obits faithfully, snip and save them, stand in the back of the old theater, feeling that warm and special glow that comes from contemplating and appreciating what has just left the building forever. The glint of gold in the sun as the newly dead lift off in their chariots. Who was Hunter S. Thompson in January 2005? A curiosity, an old writer who had made a splash as a gonzo journalist and then removed himself to a bunker in Aspen, where he fired automatic weapons for amusement; his byline, once trumpeted in *Rolling Stone* and *Esquire* and *Playboy*, was turning up on the ESPN website. But a month later, after he killed himself at sixty-seven, he was a pioneer journalist and stylistic genius who had acted as the conscience of his generation; he was irreplaceable. Legions of admirers and old friends came forth to bear witness to his legend. The old stories of Hunter streaking across the Las Vegas desert with a cache of drugs and his 300-pound Samoan lawyer, or spitting on Nixon's grave, began to glitter. These stories had been there in January; three full-blown biographies of the man had been published in his lifetime, and both Johnny Depp and Bill Murray had played him in movies. But he had slipped below the radar of appreciation, until his appearance on the obituary page gave these stories a boost. His old books began selling like mad. Mortal Hunter had ended; immortal Hunter could begin.

Death is an unstable element. Will it loft someone's reputation, or pull the ripcord? Read the obit to find out. The first obit I clipped was Rock Hudson's in 1985. *People* splashed him on the cover and told his real story inside for the first time: the romantic leading man who came out months before his death to acknowledge that he had AIDS, had lived the off-screen life of an exuberant gay hedonist. The obit as news, the obit as scoop, the obit that tells the stories that couldn't be told while the subject was still in a litigious state. Rock Hudson died—and it was so bracing to read something real instead of the hagiography that had been passing as celebrity obit, the fawning Hollywood tribute seemed to die as well. Rex Reed now recaps the year in obituaries, tattling on Janet Leigh ("I remember lunching with her once when she dropped her fork in the middle of a morsel and cried out with anguish: 'I just swallowed two raisins instead of one!'") and Julia Child ("Julia ate everything from grits to goujonettes. When the crumbs fell on her kitchen floor, she ate them, too.") Meow!

I meet the other readers of obituaries, glancing over their shoulders at the coffee shop, striking up conversations. "You like obits?" "I love them. I read them every day. I've always read them." Fervent, with a kind of feverish look in their eyes, or reverent, members of the Church of Obituaries. Some of these people read the death notices, those tiny ads paid for by the family and written by the morticians, every single day. That's hardcore. I skip those, mainly, and head right for the cream, the big feature stories of the newly departed, written by professional writers and festooned with photos. Not every newspaper lavishes space and staff writers on feature-style stories of the dead, but the ones that do—especially the big three in the United States, the *New York Times*, the *Washington Post*, and the *Los Angeles Times*, and the big four in London, the *Daily Telegraph*, the *Guardian*, the *Independent*, and *The Times*—are available on the Internet, on their own websites or in data banks. They hang on racks in the library, or pop up on newsstands, or get passed out by flight attendants. These aren't the only papers that make a big production of their obituaries. I discover a new one every week.

Once the other fans and I start talking about obits, we can't get off the subject; it takes over every conversation. The obituary pages, it turns out, are some of the best-read pages in the newspaper. Are they as popular as the sports pages? Fans of obituaries don't foam as much at the mouth as sports fans, but the emotion is there, the tension, the entertainment, the tragedy, and the comic relief.

Did you see the obit of the undercover pharmacist? Henry Giordano was an ordinary pharmacist until he began working for the Federal Bureau of Narcotics (precursor of the DEA), where he discovered a talent for undercover work.

> *He could pose as a down-at-the-heels narcotics peddler, a flashily prosperous racketeer, a small-time gambler, an escaped convict or a sailor and get away with it.*
>
> *He won the confidence of some notoriously ruthless criminal*

*gangs and helped win convictions of ring leaders and hench-
men. His work sent scores to prison.*

(*by Wolfgang Saxon*, New York Times)

In the photo, he looked like a fifties dad, receding hairline,
thick black glasses, but his bland exterior was a cover for the
James Bond of controlled substances.

How about Harold von Braunhut, the genius behind sea
monkeys? Sea Monkeys, mail-order packets of brine shrimp
that could be shipped and shelved in a dried form, sprang to
life when dropped in water; 400 million of them once shot
into space with an astronaut. I learned this on the obits
page. War, pestilence, bad investment news, and political
rants in sections A through D, but there, on the page
marked Obituaries—sea monkeys! This is the page where the
pioneers still roam, the "Pioneer of Truckers' Songs" and the
"Pioneer of Surf Music," the "Pioneer in Fighting Foot-and-
Mouth," and the "Pioneer in Frozen Vegetables." Something's not quite right when someone dies who merely "Put
Fresh Popcorn in Theaters." (Perhaps "Pioneer" hadn't fit in
the headline?)

The obit lovers I meet in the neighborhood are scrapbook
level, or below. They enjoy the stories and appreciate the
drama and humor. Maybe they save their favorites in an
album on the coffee table, their collection of "pioneer" headlines or funny coincidences, for those occasions when another
hobbyist drops by. Generally speaking, they're fans content
with one newspaper a day—two or three or four obits, a
steady drip. One of the great things about this avocation,
though, is its expandability. What begins as an occasional de-

sire to read more or better obits, or to write your own for your loved ones or yourself, can take you to the heroin level in no time.

I suppose I should stand here as a horrible warning. A few boxes of clippings, a few newspaper subscriptions—before long, I gave up my day job interviewing celebrities and spent my time hiding in a darkened room, writing their obits instead. I became popular among obit lovers, but I was socially unfit. An afternoon tea, dominated by conversations about obituaries, turned a young pregnant woman pale and then queasy, and I realized, apologizing to my hostess on the way out the door, that this was not an isolated event. I wore out my welcome visiting a friend recuperating in the hospital as well. Was there a way to talk about my work without carrying on about the dead? And you don't really want to carry on about the dead with someone who has just narrowly escaped becoming an obit subject himself. How do I say this? I scare nurses. My children are used to it, but fewer of their friends drop by, I've noticed.

A dear friend tried to talk to me about it, heart to heart. Did this obsession have anything to do with the deaths I'd experienced firsthand, a brother in childhood, close friends, family friends, the usual plume of ancestors? Of course it did. Why does someone start to read obituaries? She knows dead people. She gets older; she knows more dead people. Obituaries have a pull, a natural gravity, for those of us who've observed that life has a way of ending. But however morbidly we arrived at this page, we've ended up sticking around, hanging out, admiring the writing, getting hooked on the daily rush.

Many of us were English majors. We see this emotionally

charged block of text that follows a particular format: a swift, economical description of the person who died, a few short stories from the life or work, and the list of survivors trailing behind. This tight little coil of biography with its literary flourishes reminds us of a poem. Certainly it contains the most creative writing in journalism.

Like poetry, obituaries have had their flowery period and their bleak period and their modern period. There was the florid nineteenth century, with lots of gruesome descriptions of death scenes. "Within the short period of a year she was a bride, a beloved wife and companion, a mother, a corpse!" Things were bleak in the twentieth century; perhaps it was all those wars. A few stray writers and the occasional publication tried to jazz up the obit, working up something more than lists of white men, their jobs and clubs and descendants, but it was a gray and dusty and depressing beat for years. It wasn't just the subjects who were dead, but the prose, too. The ground began to shake in the 1980s. The equivalents of Elvis and the Beatles rose up in the United States and the United Kingdom to write the modern obituary—to give the Dead Beat a beat. Our own glorious era has been a time of expansion, innovation, entertainment, and world-class one-upmanship. In one generation, a boring, moldy old form has sprung to life. Historians tell us we are living in the Golden Age of the Obituary.

From an unsigned obit of Chet Baker in *The Times* of London, 1988:

There were certainly off-nights, but even when his trumpet tone was practically transparent, his singing voice a whisper, and the music seemingly in imminent danger of coming to an absolute halt, his innate musicianship could still achieve small miracles of wounded grace.

From Jim Nicholson's obit of "Big John" Ellis in the *Philadelphia Daily News*, 1992:

His pet saying to friends was, "Shut up, fool." Shirley Ellis hadn't been married to Big John long before she realized he was special. "We couldn't go anywhere. In New York, he'd holler, 'Shut up, fool,' and someone across the street would holler back, 'Hey, Big John.' It was that way in Washington, Wildwood, Atlantic City."

Or from an unsigned obit of Anna Haymart, who wrote novels under the name Alice Thomas Ellis, in the London *Daily Telegraph*, 2005:

Anna . . . is dressed every day in black, like some Mediterranean matriarch, from head to toe, even black tights and black flip-flops. As we sprawl over dinner, boozing and arguing like undergraduates, she leans silent against the door frame, never eating, smoking incessantly, watching, her eyes glittering in the shadows, haunting the doorway, the Woman in Black, watching.

There! A whiff of life on the page.

Across the U.S., a hybrid obituary, a cross between short stories and obits, celebrates the life of local characters, the extraordinary in the ordinary person. The school lunch lady, who spent her evenings as a ballroom hostess. The man who could hypnotize lobsters and stand them on their head.

In the U.K., four national newspapers vie every day for the most vivid, historically savvy, and gossipy obits, and have been doing so since a circulation war in the mid-1980s turned obituary writing into a competitive sport. These obits are, at their best, a form of literature, cast with the sort of characters you might find in a picaresque novel. The queen of Albania, a woman named Susan who enjoyed housewifery. The biographer who "with his alert, bright eyes . . . resembled Rat or Mole at their most benign and engaging."

The Web itself is crawling with obits and obit sites. "I walk the dead beat" is the expression; "I surf the dead beat" would be more accurate. A few keystrokes, and I can travel, for instance, the length of California, mourning the old hippies of Marin County, the artists of San Francisco, the bus drivers and hairdressers of San Jose, all the way down through L.A.'s old movie stars to Orange County's surfers and San Diego's soldiers. Obit pages cover an ever-widening pool of lives. People are still being sent off "to sing in Jesus' choir," but a growing number of journalists have absorbed the art of the obit and are writing biographical gems. Even the folksy writing has acquired a sophisticated spin.

It's the best time ever to read obituaries, and I'm here to tell you, it's a great time to die.

2

A Wake of Obituarists

Who writes obits?

People with nicknames like Dr. Death, the Angel of Death, the Doyenne of Death. A colorful nickname won't help you identify them, though. The writers called Dr. Death (and there are many) aren't wearing black robes and carrying scythes; the Doyenne of Death doesn't have teardrops tattooed down her cheek. One obit writer is nicknamed Black Mariah, after the paddy wagon that used to cart off the newly dead to be robbed before burial. This might suggest someone with a ghoulish cast, but Heather Lende, a.k.a. Black Mariah, is a lovely, cheerful woman, who looks like a model for outdoor gear. She writes simply but vividly for the weekly *Chilkat Valley News* about the people who live and die in Haines, Alaska, from the old hippies to the Tlingit natives. She talks to everybody: her book is called *If You Lived Here, I'd Know Your Name*. "In this town," she told me, "when someone dies, everyone helps—you know, brings a casserole, or offers to make the program, or play the piano at the service, whatever—and I can write." Her obits are a kind of covered dish, offerings for her friends and neighbors. When she hears someone has died, she hops on her bicycle or trudges through snow drifts to interview the survivors, who are her neighbors and friends.

Lende is the exception. Most obit writers never meet their subjects, and if they talk to the survivors, or talk to anyone, they use the telephone. It's easier to talk about death on the telephone; there's less weeping. Consequently, obituary writers are desk workers, the most disembodied contributors to newspapers and magazines. They never have to see another person; other people never have to see them. A telephone and a computer are all that's necessary to practice this craft. To the world at large, obit writers exist as bylines, unless they write for a publication that doesn't credit individual writers. Then they don't appear even on that brief line.

Obit writers are invisible, writing about people who have ceased to exist. If they've written advance obituaries for some publication's files, and they die before their subjects, they become people who no longer exist writing about people who no longer exist. This is ghostly work.

My friends laughed when I told them where I was going, the Sixth Great Obituary Writers' International Conference. And it did seem a little silly—but where else would a group of obit writers materialize? I'd find them at the conference in Las Vegas, New Mexico, in an old hotel, at the corner of a ragged historic plaza where condemned men used to be hung, in a town with lots and lots of haunted-looking real estate for sale. The Plaza, an elegant reclaimed hotel with brick walls and lace-covered casement windows, is like an old movie set for a mock western, with a hearse parked at the post in front instead of a horse. (The hearse is the ride of two of the conferees, an attractive couple, obit fans who drove up from Yuma.) Crumpled-up paper blows across the bare spots of earth, past the gazebo and into the path of the motorcycles that roar around and around the plaza, driven by tattooed and helmet-free locals. Sitting in the bar, watching through beautiful plate-glass windows, the organizer of the convention cracks over her piña colada as the motorcycles circle and growl: "You know what doctors call them, don't you? Donors."

The organizer is Carolyn Gilbert, a sixty-three-year-old public policy consultant from Dallas. She is not an obit writer, though she has written the occasional one for friends and family. Nor does she require the members of the International Association of Obituarists to be writers of obits; you can join even if you just like to read them. The organization and its annual conference sprang from a group of her friends who met regularly at Sam's, a Dallas bar. (Sam's is closed; they now meet at Sevy's.) "We likened it to Dorothy Parker's salon, but it always seemed to come back to, 'Did you see the obituary page today?'" Gilbert, a former English teacher who subscribes to a number of newspapers, including some from her old hometowns, was particularly keen on the way obits were written. They were each a revelation, even if you knew the person described. And she was impressed as she began reading more widely by all the variety and sophistication of technique. Perhaps under the influence, one of the participants suggested Gilbert organize a conference for the writers. "I'd done conferences for lawyers, state realtors—hell, I've done

conferences for every group I can think of. It was a dare. I organized this conference on a dare."

She called it the First Great Obituary Writers' Conference
as a joke, but because she is both funny and serious, she invited the professional writers and editors that she admired to
come speak. Nobody got paid for making a presentation, and
probably nobody ever will. But she built it, and they came.
The first year, in Archer City, Texas, she had twenty participants from Texas papers, from the *San Antonio Express Light* to
the *Houston Chronicle*. By the third year she moved the conference to Las Vegas, where a friend had refurbished this atmo-
spheric hotel, and it's a measure of her success that she has
repeatedly persuaded people from England and Australia to
pay their own way to a little town in New Mexico that doesn't
even have an airport.

There are fifty conferees I count at the opening reception,
and half of them write obits. I think this may be the highest
concentration of obituary writers in the world, possibly in history. This is both a thrilling and a terrifying thought. We are
tempting the fates. What if somebody unexpected dies, the
equivalent of Princess Diana, say, and all these obit writers are
stuck here in the middle of New Mexico? What if some natural disaster occurs, a fire or an earthquake, and swallows them
all up? *Dozens of obituary writers died today when a motorcycle
crashed into a hotel bar* . . . People wouldn't know whether to
laugh or cry. Across the world, reporters would have to make
something of the death of so many obit writers, and I can picture them scratching their heads, wondering, What *is* the correct collective noun for obituarists? A plague of obituarists? A
wake of obituarists?

•   •   •

Obituaries are booming as literature and folk art across the English-speaking world, and the accents that drift through the lobby the first night reflect this: southern drawls and flat Canadian vowels and British accents. Last year, several wild Australians livened up the gathering, but they didn't show this year, which disappoints the party-lovers. Jammed into the brick-lined conference room and spilling into the bar, writers and readers embrace old friends, or bend close to read each other's name tags, looking for familiar bylines.

Andrew McKie is the most visible of the U.K. contingent in his big black cowboy hat, bought for a previous conference, trailing smoke from a hand-rolled cigarette out of one side of his mouth and a ready stream of quotes and quips out of the other. He's taking a break from his job writing and editing the vastly entertaining obits at the *Daily Telegraph* to set up a post at the Plaza's saloon. In person as in print, McKie stands for pith, for the drop-dead line ("Mr. S. did not die. We regret the error and apologize for any inconvenience"), hopefully accompanied by the hoisting of a pint of something alcoholic.

McKie was only fifteen when the first wave of refreshingly frank obits hit London in 1986; he carries on a tradition handed down by one flamboyant *Telegraph* editor after another, beginning with Hugh Massingberd, the father of the *Telegraph* obit, who published such startling fare as:

The 3rd Lord Moynihan, who has died in Manila, aged 55, provided, through his character and career, ample ammunition for critics of the hereditary principle.

His chief occupations were bongo-drummer, confidence trick-
ster, brothel-keeper, drug-smuggler and police informer. . . .

McKie, a Scotsman and a showman, relishes sharing his
love of the rude obituary, and will recite without prompting
his favorite opening of all time: "Tiny Tim, the American pop
singer who has died aged sixty-two, specialized in horrendous
falsetto vocalizations of sentimental songs, and cultivated an
appearance of utter ghastliness to match." Then McKie
screws up his face for the bonus: "Can you imagine, he died at
a ukulele festival in New Hampshire!" He and his compatriot
Tim Bullamore, a freelance obits writer from Bath, are a con-
tinual source of entertaining quotes ("Dr. Atkins, the diet
doctor who did so much to help women, and blighted the
lives of their husbands and partners, died when he slipped on
an icy pavement, and couldn't get up. Why couldn't he get
up? Because he weighed three hundred pounds!") "The
boys," as Gilbert calls them, are unofficially in charge of hi-
larity. They cheerfully describe themselves as "hacks." In the
bar, rocking with the tattooed bikers and a band, they joke
about a horse some townies have tethered to the bandstand in
the plaza. One writer stayed up late enough to watch McKie
drink a forty-ounce glass of margarita, then stagger out to the
bandstand. "'I kissed that horse for fifteen minutes,'" McKie
said, or so it was reported.

He'll make a great obit someday.

Around the long tables in the conference room, after a night
of horse-kissing and motorcycle-dodging, the obit lovers sit

in semi-alert pods on the second day of the gathering, and Carolyn Gilbert begins to pluck her speakers from the audience. Half the participants seem to have spots on the agenda. Gayle Ronan Sims, the elegant lead obit writer of the *Philadelphia Inquirer*, talks about the trend of people writing their own obituaries, and asks for a show of hands: Who among us has taken care of the inevitable? "Why haven't you written your own obits?" she scolds.

From the opening remarks, I feel this autofocus zoom in as soon as one of the obituarists begins speaking; I'm enthralled by the fusion of the literary arts, black humor, and pathos, and looking around the room I see the same intensity on the faces of the others. A year later, I heard Massingberd talk about the hours he spent around the computer at the *Telegraph* with his staff, collaborating on one of their wild obits: "The favorite moment of the day, sitting around the cauldron, the computer, like a lot of witches and cackling as one was putting in the eye of newt and the toe of toad"—and I recognized the feeling.

I am an outsider only briefly. It puzzles me to see Gilbert putting her hand over her heart and turning soft whenever the name Richard Pearson comes up, and the others here seem to share this reverence. I wonder, who is this man Pearson? I have to wait only until the young obit writer from the *Washington Post*, Adam Bernstein, delivers a warm eulogy to his mentor, Pearson, the longtime obits editor of the *Post* who died this past year at age fifty-four. A man with a subversive spirit and a body that got away from him, Pearson refused to travel—except to these obits conferences. His presence lent a certain legitimacy to the ragtag gathering of scrapbookers and writers organized on a barroom dare. The man had been both a highly regarded pro and a character. He used to keep a list of cowboys and their horses' names in his desk drawer and was always writing the Associated Press wire-service editors to correct the horses' names. When someone would complain about a meager day on the obits page, Pearson would shrug and say, "God is my assignment editor," a line that now hangs over the cubicles of countless obits writers. He wrote informed and lively salutes to diplomats and world leaders, and, as an admirer of the London school of obituaries, delighted in dropping in oddball twists. His last obit, of Idi Amin, the murderous ex-dictator of Uganda, had been a romp, complete with tales of Amin carrying on dinner conversations with the frozen heads of his victims.

I chase Bernstein out to the lobby. He is so boyish and enthusiastic, I think he's a kid working his first job, but he is twenty-nine, and the *Post* is his third stop. Tall and skinny, with glasses and a prominent Adam's apple, the kid drinks martinis, listens to jazz (Artie Shaw is his hero), and reveres the great obit writers. When I ask him what made him want to be an obituary writer, he starts chanting about the big-band trumpeter who named "the zoot suit with the reet pleat, the reave sleeve, the ripe stripe, the stuff cuff, and the drape shape that was the stage rage during the boogie-woogie rhyme time of the early 1940s." I recognize this from the lead of a famous obit written by the late, great *New York Times* writer, Robert McG. Thomas, Jr.; Bernstein's mother had clipped and sent it to him while he was doing a summer internship at a newspaper in Bakersfield. "It blew my mind," he says. "I never knew you could have that much fun, not just with obits, but with any article." It gave him ideas. He volunteered to write an obit for a Bakersfield oil field worker named Bowden, called White Shoes Bowden, he learned, because the man could go into the oil fields without getting mud on his shoes. Bernstein took "a playful approach and the feedback was striking. Maybe there's a future in the past, I thought."

The *Post* has one of the larger obit staffs, and in addition to the news obituaries and columns of garish, photo-cluttered, paid death notices, also runs a weekly obit feature called "A Local Life." It's a prestigious page, and yet acclaim follows the news reporters, not the obituarists, even if the reporters are covering the new dog at the White House and the obituary writers are burying an era. "It was so perplexing to me to have these feelings. It was just like unrequited love, really," Bernstein says. But he has found his people here.

Most of his people are in their fifties, the prime age for the obits passion, and around the age that experienced journalists usually find their way back to the obits desk. Picture them as high school social studies teachers, knowledgeable and deeply engaged in current events but not show-offs. They don't tend to draw attention to themselves, in print or in person, but their obituaries are peppered with colorful tales of contemporary history and nifty turns of phrase. One of the people here writes for a tiny paper in a town with one traffic light; one writes for the *San Francisco Chronicle*. Some write obits that make us laugh and some write tear-jerkers.

One of the heart-tugging writers takes the floor that afternoon and passes out copies of a series she wrote. Trying to expand the boundaries of the form, she followed a forty-five-year-old woman dying of liver cancer and wrote about her and her family over the course of several months. I scan the headlines: *Simple pleasures gone. Telling the children. "My angel." Putting up a good fight. Not ready to let her go. A life close to ending. Mom loses struggle with cancer. Coping with her loss. Family braves first yule without mom.* Readers cried when her death finally came, and flooded the paper with heartbroken letters, and the articles won a Press Club award, so what do I know—but that poor woman: trailed by an obituarist, then battered by clichés. As I glance around the conference room, I see some of the others squirming, too, and I get a feeling for this fault line in the obituary world: news-obit writers on one side, working with hard facts, straight sentences, and a detached style about people who lived public lives, and those who write about ordinary men and women in a more personal and emotional way.

I'd love to see a match between writers of both of these styles, a cross between a debate and a rap. What if two of them really cut loose? Who would be most fun to watch? I like the former social worker who writes kindly about the old hippies who wash out in Marin County, California. Larken Bradley, who writes for *the Point Reyes Light*, leads a discussion about the emotional cost of reporting on the dead. "Does the work get to you?" she wonders, soliciting stories of writers choking up during their interviews with survivors. She has warmth and soul, and it pains her to report the hard facts of some of her subjects' lives, but she does. I also like Caroline Richmond, the tough-skinned Brit who wears giant red eyeglasses that spell LOOK, is recovering from two kinds of lymphoma ("I'm a lymphomaniac," she jokes) and specializes in writing obits of prominent doctors for the *British Medical Journal*. Given the "bloody painful" treatments she has endured, she feels that writing prickly obits of medical doctors is wonderful revenge. Bradley and Richmond represent two poles of the obituary world, empathy versus judgment, a local life versus the international stage, and though they actually coexist peacefully (and for all I know are friends), I imagine them locked in combat, firing off characteristic lines. "He was sweet when drunk," one says, pinning her opponent with a hug. "He was a snake-oil salesman!" the other retorts, flinging her into the ropes.

But none of the debates get physical. For two days, we sit civilly in a conference room in New Mexico, with oil portraits in Fauvist streaks of yellow, purple, and green gazing down at us ("visual obituaries," the artist calls them). Between reports from the professional writers, the fans in attendance pop up, one to read a funny, folksy obit from a small-town Georgia paper, another to circulate the 4,000-year-old obit of an Egyptian pharaoh—little bites of bread to clear the palate. The writers listen respectfully. These are their readers, after all. As Alana Baranick of the *Plain Dealer* puts it, the obituary enthusiasts at the conference make the writers "feel like rock stars."

For those who rise early and can find the out-of-town papers, June 5th, the last day, is not shabby as obits go, though nothing rivals the recent send-off to Alberta Martin, the last surviving Confederate widow who died a week earlier on Memorial Day and was saluted in American and British papers alike. Martin had married an eighty-one-year-old veteran for his benefits when she was a dewy twenty-one; after his death, she married his grandson. Martin's story was good for numerous follow-ups, as previously unknown Confederate widows crawled out of their closets to complain that, no, *they* were the last of the breed.

Apparently, exaggerated or even faked military service records are a real problem for obit writers. Trudi Hahn's talk "Dad served with Patton—or did he?" is one of the most popular. Hahn, the obit writer from the *Star Tribune* in Minneapolis, shrugs on a Vietnam War–era fatigue jacket dripping with medals for visual effect. At this moment in history, more than a thousand World War II vets die every day, and obit writers are ushering out the last old soldiers of the Great War; a familiarity with military matters is an essential requirement of the job. Hahn is a former Vietnam War protester who discovered a passion for covering old soldiers, dead soldiers, and POWs. She spends her vacations touring military museums and monuments, and her enthusiasm is contagious.

"This is combat insignia—it's upside down. These are corporal stripes, and they're upside down," Hahn says, pointing to the display on her chest and recommending the massive reference book *Wear and Appearance of Army Uniforms and Insignia* for all of our shelves. "This oak-leaf insignia is from WWI, but it's tilted the wrong way. This is a Distinguished Service Medal—it's given to people like Eisenhower and Patton at the end of their careers. This is a ribbon for the winner of the Pinebox Derby. And anyone can buy any of these medals on the Internet."

There are packs of obituary fans roaming the Web, or so the next speaker would have us believe. Amelia Rosner, an advertising copywriter, devotes hours every day posting to an

Internet news group dedicated to obituaries. She's not an obit writer. She describes herself as a lunatic, though she looks like (and is) a Jewish mother. Every day, she says, she reads obits from around the world, and posts her favorites on the Google newsgroup alt.obituaries. She is a particular fan of the British papers, and like many of the conferees, feels the London papers publish the crème de la crème of the form. She has them all bookmarked on her browser. Some days she posts a dozen, cut and pasted from websites and databases. "It's a hobby, but it's a very, very serious hobby," she says.

Rosner sketches out the basics of life on alt.obituaries. "Is anything sacred?" she asks the room rhetorically. "No . . . ." Last year, the members of her group spent forty-eight hours discussing people who had killed themselves on live TV. The group is particularly interested in covering the obituary in all its stages, from monitoring the celebrities who are near death to debating who should be honored in the Emmy and Oscar memorials. They track the gossip, too. "We know who's sick and who's dying, and we know how many of the original cast members of *Gilligan's Island* are still alive."

While she tells us about these super-readers and fans who gather on alt.obituaries, Rosner's online group is busy posting obits for Frances Shand Kydd and for the man who survived 9/11 only to be fatally clubbed outside of Madison Square Garden after a Who concert.

Also posted this day, one of the occasional "Healthwatch" notices that crop up when someone is hospitalized or appears in public looking particularly pale. In this case, a CNN story is quoted that claims "Former President Ronald Reagan's health is deteriorating." Unnamed sources at the White

House said they were told: "Don't be surprised if . . . the time is getting close."

The gathering of obituarists draws to an end. Andrew McKie ends his rollicking talk with the line, "We all know this is the only job," and the room cheers. One last presentation, a slide show on Bath, England, where the next Annual Obituary Writers Conference is slated to take place, and it's over.

There's a short break while the slide show is being rigged. Rosner and her pal, Stephen Miller, who writes for the *New York Sun*, step out to the lobby to check up on alt.obits on the courtesy computer. The next presenters fuss over the slide projector loaded with pictures of spas and cobblestoned streets and tea shops a world away.

Suddenly, there is a commotion, and Rosner and Miller come running back. "Stop the presses!" they shout (a quaint expression from the great age of newspapers). Rosner is in her early fifties, and Miller is ten years younger, but they look like children with flushed cheeks and open mouths. There are people in the room, including McKie, who instantly divine the news and start crowing, but a few of us spend a half-minute in confusion. Someone is dead, someone big—who? who? Reagan?

Not Reagan!

Yes, Ronald Reagan!

"Oh, my God, Reagan's dead!" "No!" "He died at the end of the obituary conference?" "Can you believe his timing?" "Wild, isn't it?"

One reporter grins and says, "How do I describe this

room? It's electrified!" I came to see what obit writers look like, and this is what they look like: regular people, who happen to spring to life when bad news arrives.

This conference had been haunted by previous years; too bad you missed the wonderful Myrna Oliver from the *Los Angeles Times*, I've been told; or you should have been here with Dr. Nigel Starck, who has been crawling the globe for years for his history of obituaries. With one well-timed death, though, this has become the enviable year.

"I just had a hunch," Rosner says later. "There had been rumors for days," Miller adds.

The photo editor from the *Dallas Morning News* is dismayed—here she is, in the middle of nowhere, and the biggest story in her files is going to run without her supervision. McKie finds the single pay phone in the lobby being hogged by the Kid from the *Post* (who doesn't even need to check in with his paper—the *Post* was ready with its Reagan obit when Bernstein was still in college; "I called to find out what was happening back in the newsroom," he explains, but he's seen *The Front Page* and always wanted to hang an OUT OF ORDER sign on a phone to beat out the competition). McKie says, "I had to run up to my bloody room," and returns, groaning. "It's ten at night in London, and there are two people there running around like headless chickens." He has a long file in the office computer ready to go that the news reporters will plunder. "I hope they don't nick too much from my queue."

Stephen Miller shakes his head. He has only one advance obit on file, and it's not for Reagan—it's for the pope. There's nobody covering for him at the *Sun*. His editor will have to reprint the story from the Associated Press.

Before we break, about as pumped up as a room full of writers can get, Carolyn Gilbert succumbs to nostalgia. There have been a lot of romances at the conference over six years, resulting in three marriages but she announces with satisfaction, "This is the only death. Bob Hope died during the second conference, we thought—but then he rallied."

We may be in the wrong place, but we're *in place*. Obituarists wait for death every day, and Reagan, who had been shot once in an assassination attempt, was ninety-one and had Alzheimer's; he's been missing from public life for years. There is a kind of completion in his death, a harmonic aspect to its timing.

A few of the writers hurry to catch a plane out of Santa Fe; the rest turn the lively bar raucous. The other event of the day is happening at the Belmont Stakes, where Smarty Jones is getting ready to run for the Triple Crown (and where we learn later, the announcement of Reagan's death to the clubhouse began soberly, "Ladies and gentlemen . . ." horrifying the racing fans, who thought something had happened to the horse.) Smarty Jones loses to a forgettable steed. The adrenaline dissipates. At five A.M., dragging my suitcase through the lobby, I find only Trudi Hahn left, tapping on the computer, bingeing on the early Reagan obits. Farewell to the wake of obituarists.

The conference, with its improbable but perfect eleventh-hour death, is written up from London to Toronto, and New York to Dallas. Rosner posts Reagan obits from around the world, as well as an account of the conference, on alt.obituaries. The Kid covers the story on National Public Radio.

They all make the same point, more or less: *Forgive us, but this is what we live for.*

3

Name That Bit

There's a question, a techni-
cal matter, which has troubled me since I started reading obituaries: What do writers call *that phrase*, the telling clause that
appears in the first sentence of almost every obit? It's the dis-
tillation of all the departed has meant to us—the obit within
the obit. Robert McG. Thomas, Jr., of the *New York Times*
happened to be the master of *that phrase*.

John Fulton, a Philadelphia-born artist who worked with
cape and sword in the bullrings of Spain, then celebrated
his momentary masterpieces of ritual death by painting
pictures of the very bulls he had slain using their own
blood, *died on Friday* . . .

Edward Lowe, whose accidental discovery of a product he
called Kitty Litter made cats more welcome household
company and created a half-billion-dollar industry,
died . . .

Elizabeth Bottomley Noyce, a microchip millionaire's scorned first wife who showed as much imagination and verve in deploying her half of his Silicon Valley fortune as he had in making it . . .

The only label I have found was coined by James Carville, the firebrand political consultant; he referred to it as *the comma*. Not bad, but it's such an important part of the obit, doesn't it deserve something dashing and slangy from the world of newspapers? At the obits conference, Andrew McKie spoke about an interesting difference between American and British obituaries. The point of the American obit is the death.

Billy Carter, the irrepressible gas station proprietor and farmer who vaulted to national celebrity in his brother Jimmy's successful campaign for President in 1976, died of cancer of the pancreas yesterday.

The Brits prefer to bury the news of the death in the middle; for them, the death is simply an excuse to write about the life.

Billy Carter, who has died aged 51, was President Jimmy Carter's hard-drinking roly-poly brother whose bibulous verandah-chair comments from the peanut township of Plains, Georgia, caused periodic embarrassment at the White House.

Here was an opportunity to find out what vocabulary the obit professionals used for the thumbnail description, wherever it appeared. I raised my hand for McKie. What do you call the clause that follows the person's name and sums him up? I asked. He looked at me blankly. You know, I said, *that phrase*. How do the people in your newsroom refer to it? It's an adjectival clause, he said. Well, I know that, I argued, feeling let down, but doesn't it have a special name in the obituary form? "I call it, I call it . . ." McKie sputtered, jabbing his forefinger to a page. "I call it *This bit here!*"

The obits editor of the *New York Times* calls it "the who clause." The editors of the *Dallas Morning News* call it "the descriptive." *Philadelphia Daily News* writer Jim Nicholson, when asked if he had names for the conventional parts of the obits he wrote, particularly the subordinate clause that follows the name and precedes "died on Friday," replied, "Subordinate clause? I wouldn't know one if it bit me in the ass." He said he almost failed English when grammar was on the syllabus. "There was never a name for it, nor was any format ever suggested or imposed on me."

But there is a format, a template for the obituary that almost every newspaper publication follows. Writers have absorbed it, and readers come to expect it, almost as if an invisible metronome accompanied its unfolding. Writers and readers have internalized the form, just as we've internalized the rhymes and rhythm of a limerick, or the structure of a joke. But though names exist for the various components and literary devices of poetry and jokes (stanza, slant rhyme, punch line), the obituary's unique and particular elements are anonymous little tools that can be described only with effort. If I want to refer you to one of the obituary's standard devices, for instance, I have to send you to the shed with a clunky and possibly inchoate description—that tool with the sharp prongs that you use to make a little dig at the subject . . . a cultivator? We're in uncharted territory, rooting around the obituary shed. Where are the Ph.D.'s when you need them?

I'll have to invent the vocabulary myself. As I write this, it is St. Patrick's Day, and a good one for obits, primarily because the *Telegraph* has a corker about the cross-dressing performer who fought for Hitler, had a sex change, and found fulfillment as a professional whistler in Austria. The only aborigine judge in Australia died, as did a woman who spent fifty-one years as a mental patient in Texas because her parents didn't want her to go dancing. I'll use them and the other obits of the day to illustrate my modest proposed lexicon.

Let's start with the bit I've been calling *that phrase*. Here's the way *that phrase* appears in one of the British obits:

> *Jeanette Schmid*, the professional whistler who *has died in Vienna aged 80*, performed with Frank Sinatra, Edith Piaf and Marlene Dietrich; she had been born a man and had fought in Hitler's Wehrmacht before undergoing a sex change in a Cairo clinic.
>
> (*Daily Telegraph*)

It's slightly complicated, but worth it; like a sparkler, it shoots off luminous details in all directions. "The professional whistler" alone would make the obit interesting, but to also have Hitler, Frank Sinatra, and a sex-change operation *(in a Cairo clinic!)* is outrageous. The business about dying "aged 80" sits in the middle of the sentence, British style.

In the American model:

James R. Garfield II, father of the modern Cleveland auto show and great-grandson of an American president, *died of a heart attack Tuesday at LakeWest Hospital in Willoughby.*

(*by Alana Baranick*, Plain Dealer)

God bless Alana for that "father of the modern Cleveland auto show." It's priceless! But what is it? Is it an *appositive*, a grammatical label for a lovely, side-by-side pair of nouns that have the same referent? Not quite, and anyway, that's not fun enough, and we need fun labels for the obituary's pieces. In the phrases above, the descriptions are a bit more elaborate than what might appear carved on a tombstone, but it's the same idea, isn't it? An economical point or two designed to give the deceased a memorable label. So how about *the tombstone*? Picture a harried editor saying, "I've got the subject and the tombstone, can somebody else go wrap it up?" The part that's most important has been chiseled in stone.

What follows *the tombstone* is usually a report of the circumstances that landed the person on the obits page.

Judge [Bob] Bellear died on Tuesday, aged 60, the deadly dust he inhaled over 30 years ago cutting short his life and trailblazing legal career.

(*by Kim Arlington*, Associated Press)

This element is often stripped down in the British papers to the date and location of death and the age of the deceased. They skip the cause of death if it's the usual pack of wolves, unless there's a tale to tell. The British quietly admit to being disgusted by the way the Americans wallow in medical details, and while I do like to know the particular disease or disaster that carried off the deceased, they have a point. I was horrified reading in an otherwise lovely obit in the *New York Times* that the subject, "who was recovering from a blood infection he contracted several months ago, choked to death after vomiting." Blood and vomit—wow. "Those endless paragraphs," Hugh Massingberd complains. "They go on and on until you think, 'I really can't take this anymore.'" Indeed! So what do you call that bit there? Doesn't *the bad news* describe it?

When the London papers mention the cause of death, they like it to be germane. How much more pleasant, if pleasant is the word, to see *the bad news* couched in an anecdote about the life, as the following demonstrates so gracefully:

> *At a party held at the Scottish National Portrait Gallery in Edinburgh last October to celebrate* The Biographical Dictionary of Scottish Women, *Sue Innes, historian, writer, feminist activist and a co-editor of the dictionary, acknowledged that celebration was a bit premature: the book was not scheduled to be published for at least a year. "I'm sure there will be more parties around the dictionary," she said, "but I may not be able to attend them all. I wanted us to have a party that I could attend."*
>
> *Having thus acknowledged the brain tumour which (as everyone in the room knew) was killing her, she moved off in her wheelchair to chat to friends. Five months later, she died, aged 56.*
>
> (*by Zoe Fairbairns,* Independent)

What tends to follow *the bad news* in the obituary is an expansive section of one or more paragraphs, an anecdote or even a full-blown scene that illustrates the turning point in the story of the subject's life.

On Dec. 7, 1941, when the Japanese attacked Pearl Harbor, Mr. [Charles] Cioffi raced in his boxer shorts from his bunk to the deck of the light cruiser Helena. As he fired round after round from a machine gun, a torpedo slammed into the Helena's starboard side. Mr. Cioffi, a seaman first class who would rise to the rank of chief gunner's mate, suffered severe burns that left scars on his arms, legs and torso. . . .
(by *Jack Williams*, San Diego Union Leader)

Five years ago, he chanced on a story about an effort to add military recognition to the funerals of veterans. To [David G.] Hurley's horror, as a dedicated musician, he learned that taps were frequently played from a cassette tape on a boom-box. So, he offered to travel to West Coast funerals, at his own expense, to play live taps on his trumpet. Ultimately, he coordinated a loose-knit group who played at old soldiers' funerals. Mr. Hurley himself performed at about two dozen funerals. Last summer, he was in Eureka to bugle for a former general. . . .
(by *Jane Kay*, San Francisco Chronicle)

This is the story in the story obit, the action, the glittering showstopper where the reader expects to be dazzled or transported. I call this *the song and dance*. (In the more stylistically jazzy papers, and sometimes in the hybrid American obit, *the song and dance* sometimes comes before *the tombstone*.)

In the boilerplate obit, the writer will at once drop back, take a breath, and put *the song and dance* in context. Ordinarily, this signals a recap of the résumé and lets us know that the writer is going to back and fill the story. Okay, you've had your fun, now sit still while we feed you some facts. You can't miss this shift; it almost always starts where it all began, with the birth of the subject, and it's usually more mundane and boring than the following two examples:

> *Born near Murwillumbah, on the NSW north coast, Judge Bellear joined the navy to help support his eight siblings and later worked as a fitter and turner.*
>
> 　　　　　　　　　　　(*by Kim Arlington, Associated Press*)

> *Born in rural Saskatchewan, [Lawrence Roluf] grew up in Winnipeg, and Red Lake, Ont., where his father owned the local movie theatre.*
>
> 　　　　　　　　　　　(*by Carol Cooper,* Globe and Mail)

The reader has been cruising along, following the action, and suddenly has to stop and go back. If the obit were a car, and the obit writer the driver, this would be *the reverse shift*. I like the idea of there being a set of gears in the obit that the writer has to grapple with.

And then begins the chronology of the subject's life, which, in the wrong hands, can be long and dreary. James Fergusson, the *Independent*'s obits editor, calls it *the desperate chronology*. The measure of the artistry of an obit writer is his or her ability to transcend that *desperate chronology*. Nice phrase! I vote to keep it, if only as a reminder of how difficult it is to escape.

The *Independent*, in an effort to dodge *the desperate chronology*, appends to its obituaries a little shaded box with the particulars of the subject's birth, death, marriages, and positions, and a few other papers have followed. Fergusson calls it "the endpiece." I considered calling this appendage "the megillah," as in "the whole megillah," from the Hebrew word for scroll, until I talked to Colin Haskin, the editor of a dynamite obit page for the Canadian paper the *Globe and Mail*. He calls it *the black box*, which is apt, even if the box it's in isn't black.

Sprinkled throughout the obituary, but usually appearing between the chronology and the end—the second-to-last word, as it were—are the colorful quotes, from experts, relatives, old friends. To call them colorful quotes doesn't do them justice, because they have a purpose in the obit beyond providing the splash of a living voice. They are there to establish a crucial psychology, to dig under the usual facts about a life and communicate something of the obit subject's inner tickings. Imagine a round table, and the people who knew the deceased standing up, rapping on their glasses with a spoon, and saying something that fills in the blanks, directly or indirectly.

Here's one, from the send-off to a regular guy who ran a deli in a mixed neighborhood:

"We moved to the neighborhood not just for the house, but for people of all stripes," he said. "Sid [Drazin] was his own separate stripe. He also had great bagels."

(*by Paul Schwartzman*, Washington Post)

The obit of Opal Petty, the woman who spent decades in a mental institution, was rich with loaded quotes. The director of the Texas Civil Rights Project, who filed the suit that led to her freedom, explained her story in a nutshell:

> *"Being fundamentalist Baptists, her family didn't approve of her wanting to go out dancing and such things. A church exorcism didn't work, so the family made the decision to commit [Opal Petty]. One of her girl friends said she didn't see anything wrong with her."*
>
> (by Christopher Lehmann-Haupt, New York Times)

It's almost inconceivable, this woman's life: locked up at the age of sixteen, zombified by that jail of an institution, rescued more than fifty years later by her nephew's wife, who wondered (as no one else had) what had become of her. Petty lived for almost twenty years with the nephew and his wife, working at a job for the mentally disabled and buying dolls for her collection with her paltry earnings. As the wife observed,

> *"They were her family. When she was buried she was surrounded by her dolls."*

In an obit of Van Chu, a refugee who became an advocate for refugees, his younger daughter explained why he'd moved his family out of Vietnam:

> *". . . for us to have a better life and not have to pick up garbage for Ho Chi Minh."*
>
> (by John Iwasaki, Seattle Post-Intelligencer)

These quotes sizzle when they hit the grill, but they have substance, too. A glass is raised, and someone who knew the subject (a daughter, or customer, or wife of a relative) delivers something personal and revealing. I think we should call such quotes *friars*, after the club famous for the toasts and roasts of its members. An obit without a *friar* or two is an obit without flavor or texture.

The deadpan jokes and juxtapositions that set the tone for so many obits would probably not exist were it not for the *Daily Telegraph*.

> *When her husband became Prime Minister in April 1976 following Harold Wilson's resignation, the headlines did not spare Audrey Callaghan, labelling her "the Yorkshire Pudding," ostensibly for her skill in cooking, but alluding to her poor dress sense and mildly disorganised appearance. Mrs. Callaghan, said a woman columnist, had "resolutely resisted the traditional tendency towards neat elegance and suitable hats." It was also noted that her hobby was keeping pigs.*

The drumroll builds till—crash!—both drumsticks land on the pigs. Those pigs are the jolt at the end of a series of facts, the sly wink after a string of descriptives. (The *Daily Telegraph* goes on to point out that Mrs. Callaghan's "homely exterior concealed a good brain and a strong political will," so see? They can get past the superficials on a woman.) Since this is a standard device in its obits, I propose that it be called *the telegraph*.

An advertising illustrator from age 18, his work appeared in Vogue, Sports Illustrated, Time, Life, Esquire, The New Yorker *and other magazines. He illustrated for children's books later in life. [Ted] Rand also painted portraits of dignitaries, including several members of the Saudi royal family who hung his work in airplanes. . . .*

(*Associated Press*)

I can't say whether the anonymous author of the above list knew that it would culminate in such a funny anomaly, but it doesn't matter. Something about the rhythm of these sentences begs to end in a punch line; the arc of a paragraph in numerous obituaries ends this way, on that wry note that telegraphs irony—*the telegraph.*

Mr. [Dick] Radatz spent the last two years as pitching coach for the Lynn-based North Shore Spirit, an independent minor league team, and he was planning to return this spring . . . even though Mr. Radatz's considerable girth—his weight approached 400 pounds—made trips to the mound a rarity.

(*by Gordon Edes,* Boston Globe)

There it is—"made trips to the mound a rarity." It's a zinger, delivered with a poker face. *The stinging telegraph.*

Finally, the standard obituary ends with the business of listing the survivors. "List of survivors"—that's what people call it, so of course it's pedestrian. You have to really hunt to find a "list of survivors" that doesn't have more in common with the phone book than a news story. The rare variations are welcome:

Among his survivors, Mr. Radatz had a son, Dick Jr., who runs a collegiate baseball league in Michigan, and a daughter.

Lady Callaghan is survived by her husband and their son and two daughters, one of whom, Margaret, is Lady Jay of Paddington.

Perhaps we'd see more of its dramatic potential if we called it *the lifeboat.*

4

<hr>

The Mighty and
the Fallen of New York

THE IRISH SPORTS PAGE

The tall man with the Irish face was sitting on a commuter train late in the morning, meditating on his book. He probably boarded in Katonah, close enough to the first stop to get a nice seat, mid-car, at the window of a three-seater. The trip on the Harlem line isn't the most beautiful, straight down the middle of Westchester County, not like the Hudson line along the water, where the river comes within feet of the tracks and the George Washington Bridge glitters overhead. But, still—green trees, rolling hills, granite outcroppings, along with the swingset-studded backyards and the bus yards and acres of SUVs and station wagons.

At my stop, I stepped on and popped into the end of the first long seat with only one occupant. I was on my way to interview Chuck Strum, the obits editor of the *New York Times*, one of the most powerful people in the obituary world, and I was nervous. I was together enough to be carrying notepaper, pencils, and a tape recorder, but not enough to remember the list of questions I had prepared to ask him. There was static flying off me when I realized this—if I were a cartoon character, a few squiggly lines in the air around my head would have conveyed it—and, rooting around in my purse, I dropped my cell phone onto the seat and it tumbled to the floor. My seatmate lunged, but the phone slid under the seat. He said sympathetically, "Oh that's happened to me—on airplanes, which is worse." And that was when I saw that that great stillness next to the window was the former poet laureate of the United States, Billy Collins.

He's a personable guy, with an open, friendly face; you don't need to know him to have a conversation. This is the poet who mocked the pedants' urge to "tie the poem to a chair with rope and torture a confession out of it," whose latest book is called *The Trouble with Poetry*, an informal, sociable man who could talk about anything—until the topic of death was introduced, after which he could talk only about obituaries.

He was riding in to promote a new anthology of poems he edited, *180 More*. He held it up, show-and-tell, and was pleased when I took it and started riffling through its pages—mostly living poets, though there were a handful who died in the nineties and early this century. It's not that I don't care about the living, but the year 2004 was a terrible year for poets, and sure enough that cascade of the recently dead are represented: Donald Justice and Czeslaw Milosz, who died in

August, followed by Virginia Hamilton Adair and Michael Donaghy in September, then Anthony Hecht in October, all old if not ancient, except for Donaghy, who died of a brain aneurysm at fifty. I hated seeing that fifty, my age. Poetry is not a pressured or dangerous activity in most countries. Non-suicidal poets tend to live long lives, and poets cherish their elders; they grow old and are *not* forgotten. Stanley Kunitz was about to turn one hundred, and had to run around constantly being honored. Collins himself was only sixty-three. He shook his head. Donaghy was practically a kid.

Michael Donaghy wasn't famous here. He didn't rate even a short *New York Times* obit. He was an Irish-American who emigrated to London and published three wonderful books of poetry in England, where he was known and beloved for his poetry and the Irish music he made. But anyone who had seen him read his work would never forget him. He memorized his poems and delivered them with audacity—chatty, witty poems, often rhymed, with humble subjects and beautiful phrasing: a lovely mix of the high and low arts. The audience I saw him with had been riveted. Months later his obituaries and letters from his stricken friends filled the London papers. Even the potentially snarky *Telegraph* had seen fit to write a long and literate appreciation, honoring him as a "prizewinning poet whose own stories, such as when he manhandled Pavarotti into a cab, found expression in formal, elegant verse." "Formal and elegant" doesn't quite convey it, though. Here's that poem in Collins's anthology, "Local 32B," about Donaghy's stint as a doorman, with its prophetic line, "An Irish doorman foresees his death."

Once I got a cab for Pavarotti. No kidding.
No tip, either.

The poem ends:

Yessir, I put the tenor in the vehicle.
And a mighty tight squeeze it was.

We sat on the late-morning train, laughing over the vivid voice that had survived the death of its speaker. Then Collins added a delicious footnote. "Did you know the 'tenor' and the 'vehicle' are literary terms? They're the parts of a metaphor. In 'My love is like a rose,' for instance—'my love' is the tenor and 'a rose' is the vehicle. 'I put the tenor in the vehicle.' Such a clever man."

Collins's father (and many others, no doubt) used to call obituaries "the Irish sports page." Collins missed the whimsical obits of Robert McG. Thomas, Jr.; they met his high literary standards. We rode a few minutes with our silent thoughts. Obituaries made the great sprawl of New York a small town. "You know, I once wrote a poem called 'Obituaries'" Collins said, a little shyly.

. . . eventually you may join
the crowd who turn here first to see
who has fallen in the night . . .

As if I didn't know he was one of us!

THE FRANCHISE

Chuck Strum nurses a very dry Beefeater martini with olives and patiently fields a series of questions about Alden Whitman and Robert McG. Thomas, Jr., the two star writers of the *New York Times*'s obits page who came and went before Strum became the obituary editor. Salt-and-pepper hair, a tweed jacket, absolutely precise with his information (he knows what he knows, and also what he doesn't know), and equally precise with his timing (he left detailed instructions and apologies when he thought a meeting would make him five minutes late)—Strum is both a gentleman and a pro. One thing he is not is flamboyant, which both Whitman and Thomas were. It must drive him nuts that people like me keep bringing them up.

Alden Whitman was a bow-tied Harvard graduate and former copyeditor who affected a French policeman's cape and cultivated the mystique of interviewing his prospective subjects while they were still (obviously) alive. Harry Truman enjoyed himself so much that he put Whitman in touch with the man planning his funeral. "Mr. Bad News," as Gay Talese called him, was a man with a "marvelous, magpie mind cluttered with all sorts of useless information. . . . All day long while his colleagues are running this way and that, pursuing the here and now, Whitman sits quietly at his desk near the back, sipping his tea, dwelling in this strange little world of the half-living, the half dead in this enormous place called the City Room." For about a decade, from the mid-sixties to the mid-seventies, Whitman ushered out such giants as Helen Keller, Charlie Chaplin, Picasso, and Ho Chi Minh, polishing the *Times*'s reputation for

sweeping and colorful first drafts of history. He savored the delicious detail, the ashtrays shaped like pianos in Liberace's man-
sion, Elizabeth Arden's horses in their "fly-free" stalls.

Robert McG. Thomas, Jr., on the other hand, was a rangy
Tennessee native who flunked out of Yale and became a copyboy, then a reporter, for the *Times*. He had written obits off
and on during his long career, but landed for good on the
obits page in the 1990s, in punishment, I suspect, for some
transgression—which is how many seasoned reporters used to
tumble onto the page. Thomas exercised a Southern flair for
shaggy tales about odd, inherently funny people like the
queen of chopped liver, the king of kitty litter, and the Goat
Man ("You take a fellow who looks like a goat, travels around
with goats, eats with goats, lies down among goats and smells
like a goat and it won't be long before people will be calling
him the Goat Man"—so that one began). He wrote his
loosey-goosey riffs on short deadline, and "he sometimes ran
into career turbulence because of an acknowledged tendency
to carry things like sentences, paragraphs, ideas and enthusi-
asms further than at least some editors preferred," as Michael
T. Kaufman wrote diplomatically in McG.'s obituary in 2000.
McG. made the funky, funny obit a regular part of the *Times*'s
obits page, and his influence endures: it may be a good day on
the obits page, but it's not a great day if there aren't any
"yarns," as the *Times* calls them—the longest-working vendor
at Yankee Stadium, who died at the start of the 2005 season,
or the Rothschild heiress who was also a world expert on
fleas. These yarns are a crucial part of the *New York Times*'s
obits page, a page Chuck Strum terms "this great franchise."

Strum, fifty-seven, is sitting at a red booth in a swank midtown café, swirling his gin, forking a man-sized salad, a powerful man savoring his perch. He is also sitting atop a figurative mountain papered with tens of thousands of stories of compelling and important lives, written by thousands of talented men and women over generations—the history of obits in the newspaper of record. The *Times*, he points out gently but firmly, has been publishing fascinating obits since 1851, when it saluted the Reverand Thomas H. Gallaudet, "the pioneer of Deaf-Mute Instruction in this country," and the novelist James Fenimore Cooper ("In Europe, and especially on the Continent, no American name stands higher than that of Cooper").

The *Times* prides itself—to a fault, some say—on setting the standard for American journalism, and the obits page is no exception. You may prefer those in the *Los Angeles Times* or the *Washington Post*, or be partial to the *Plain-Dealer* or the *Boston Globe*, or any of dozens of other papers in big, mid-sized, or tiny cities where good obits find a home. But it's unthinkable to skip the *New York Times*. Even if there's no McG. to look forward to, there are writers like Douglas Martin who sometimes seems to channel him:

Selma Koch, a Manhattan store owner who earned a national reputation by helping women find the right bra size, mostly through a discerning glance and never with a tape measure, died Thursday at Mount Sinai Medical Center. She was 95 and a 34B.

The *Times*'s newest regular obits writer, Margalit Fox, has been spreading her quirky wings on the filmmaker who cranked out all those classic classroom reels "from 'Squeak the Squirrel' to 'Teeth Are for Life,'" and the founder of Matchbox cars, who "for several decades after World War II was the world's largest automaker." Fox waited ages for a chance to compete for the opening on the obits page; times have changed when the seat that used to hold reporters on their way out the door is one that reporters compete for. Though there is still some stigma attached to writing obits, it's fading, Strum feels, and the number of inquiries he got when the last spot opened up reflects that. "Inevitably anyone who writes an obit for me, a daily or an advance, when it's published finds out that more people read that article than almost anything else they'd ever written."

Strum would be a happy man if he could run the equivalent of McG.'s kitty litter king every day—"Who in the United States has not been touched by kitty litter?"—but entertainment is the occasional bonus of his obits page, not its aim. The historical record is the point. Douglas Martin's vivid obit on Lisa Fittko, a World War II heroine who smuggled numerous people out of Europe, appeared nine days after her death because, Strum says, "You can't know all of this stuff. That whole period is extremely vague. There are people who will tell you they did this, that, and the other thing, and Doug took days to separate the wheat from the chaff. The *Chicago Tribune* ran an obit that then had to be corrected extensively because it was all 'ucked fup,' as they say in the business." Though Fittko's obit weaves together multiple stories and locations and mentions more than a dozen names, one peripheral name had to be corrected the next day; an *s* had been mistakenly tacked onto a French surname. Maybe there's a dungeon in the basement of the *Times* building where you get lashed for that error, but just one letter of the alphabet wrong? An "A," not an "A+."

Why are obits such a pain to report? Other reporters "follow stories that are running stories," which develop over the course of days or weeks, with denials and corrections and additional information being part of each day's update. "When you get an obit," Strum points out, "you're starting from square one." Like Trudi Hahn in her fatigue jacket dripping with fake medals, he bemoans unreliable families, who "have reason to keep information from you, or who don't remember, or who think they remember but assume certain things— they give you names wrong, they give you dates wrong, they give you information wrong, they give you history wrong, they give you things that have come down to them as a mythology—and checking that kind of material on a deadline is almost impossible." The *Times* tries to publish every correction, part of a campaign to dismantle the fortress, begun in earnest after reporter Jayson Blair embarrassed it with years of stories he made up over lunch. These detailed daily corrections make it easy to see how vulnerable the densely factual obits are to error, and Strum is surprisingly open about errors, like the still living people who appear on occasion in his pages, another professional hazard. From the summer of '04:

An obituary on this page yesterday erroneously reported the death of Katharine Sergava, a dancer and an actress who portrayed the dream-ballet version of Laurey, the heroine, in the original production of "Oklahoma!" Friends of hers reported the error yesterday.

The obituary was based on one in The Daily Telegraph of London on Nov. 29. The Times was unable to confirm her death independently and, through reporting and editing errors, omitted attribution. The Telegraph says it has begun its own inquiry.

Not long after I met him, Strum ran a correction for an obit that had appeared twelve years earlier. So what? The electronic record is now accurate. It's a more serious business than it used to be; for instance, in 1854, this early *Times* correction ran under the headline OFFICER BOWYER NOT DEAD:

A report got into circulation yesterday that Police Officer Bowyer had departed this life. On the contrary, that popular gentleman was walking the streets yesterday in perfect health. The report originated from an obituary notice, published, nobody knows why, in a morning newspaper.

As a rule, Strum avoids running obits that claim somebody was the oldest person in Japan, or the last of the Civil War widows, or the last member of Howdy Doody's cast to die; anything involving patents is cursed, as far as he's concerned, not to mention "the affection that writers have for the narrative form without attribution—it's not good." Those wonderful spinners of yarns, like McG., need an editor to rein them in.

It's a challenging section to edit, and never more so than when a famous person dies unexpectedly, like late in a three-day weekend, during a snowstorm, when there's no advance obit in the bank, and the only available writer is too young to know much about the subject—the unfortunate confluence of circumstances when word arrived that Hunter S. Thompson had shot himself. Strum spent part of the Sunday night of Presidents' Day weekend on the phone, making the case to the young writer for gonzo journalism, what it is and why it's important. To hear Strum, an old-fashioned newsman from New Jersey who has spent thirty-four years at his craft, the last twenty-six at the *Times*, expound on the merits of gonzo must have been priceless. But he appreciated Thompson as a reporter, and while other obituarists leaned heavily on Thompson's flamboyant use of drugs and guns, Strum got it. The Monday obit was cursory but respectful, and satisfied the demand for news. The ultimate obit that ran Tuesday, by Michael Slackman, reflected Strum's understanding of Thompson's significance.

> *During his career, there were moments, usually in interviews or in his own personal correspondence, when Mr. Thompson let the public in on the point. It was, he seemed to suggest, not really about guns and drugs, and tearing up the pavement and planting grass, but about grabbing public attention to focus on the failures of leadership, the hypocrisy in society. . . . "The real issue [Thompson said] was Power and Who was going to have it."*

This afternoon Strum gets to edit his favorite kind of obit, the story of a man few people have heard of but everyone has been touched by—one Maurice Hilleman, "a microbiologist who developed vaccines for mumps, measles, chickenpox, pneumonia, meningitis and other diseases, saving tens of millions of lives . . . probably more lives than any other scientist in the 20th century." People who play the horses would call this an "overlay," a great horse that most bettors would overlook. Hilleman's life is so remarkable, yet unknown to most readers, that Strum's managing editors decide to start it on the lower right side of the front page of the *Times*. After a suspenseful account of the science and artistry required to develop all these vaccines, Lawrence K. Altman, a solid medical reporter with a graceful style, draws a picture of the man himself:

> *Dr. Hilleman stood 6-foot-1, wore reading glasses that rested on the tip of his nose and described himself as a renegade. He often participated in scientific meetings, where he could be irascible while amusing his colleagues with profane asides.*

In *the reverse shift*, Altman goes back to his beginnings, writing that Hilleman "was born on Aug. 30, 1919, in Miles City, Mont. His mother and twin sister died during his birth."

Nothing could be more straightforward, the unadorned linked facts of the man's birth, but if you are reading it as Altman and Strum want you to read—that is, if you have been caught up in the story of a man responsible for the health of millions of children—the fact that he survived while his twin and his mother didn't is startling. What if Hilleman, too, had died at birth? Or, given the good he did the world, what might his sister have turned into? And how might his whole story have changed if his mother had lived?

Strum admires these straight and solid writers who find drama not in artful prose so much as scrupulous reporting and skillful juxtapositions. He mentions Robert D. McFadden, who has had a frontrow seat for more disasters than most of us remember. The great rewrite man, a Pulitzer Prize winner, has the ability, Strum says, "to take scores and scores of scraps and turn them into a beautiful, coherent story, all fully attributed. He asks the reporters who do his legwork, 'How many steps up to the apartment building where the dead man was found? Did he have on red socks or blue? Was the paint peeling off the building or was it freshly painted? How high were the houses? Were people milling about on the street or were they behind bolted doors?' And then he stitches these details together brilliantly. Like a great murder mystery, except it's all true."

Strum was tapped four years ago to fill the shoes of Marvin Siegel, who edited the *Times*'s obits as well as a collection, *The Last Word: The New York Times Book of Obituaries and Farewells, a Celebration of Unusual Lives*. Out of 130 bylines in this collection, a quarter of them are McG.'s. Strum says he was picked for Siegel's job because he knew "a little something about everything." He also has what I've come to think are the contradictions of every good obit writer: empathy and detachment; sensitivity and bluntness. Respect prevails, but it's livened by a bit of waspishness. Strum says deciding who gets obits is the hardest part of his job. "I suffer and grieve thinking we might be overlooking someone deserving," he declares. But he also calls the famous writer who begs him to run a relative's obit "a pain in the ass."

Strum would welcome a redesign and his own picture editor, and he'd like what every paper wants, more great reporters and researchers and writers and resources. He's grateful for his deputy, Claiborne Ray, a small woman with wisps of hair flying around her face, who corrals the daily hurricane into a memo of the day's dead and nearly dead. "She'll say, 'Remember Betty Hill, the woman who was abducted by aliens?'" Strum says. "She knows stuff like this, the history of everything. She's like the canary in the mine," he declares, then chases his martini with a double espresso before strolling back to Ray's memo for the day and the snowballing deadlines.

Strum's office is in the aging, creaking Times building, a kind of fortress in mid-Manhattan; there are plans to move to more modern quarters in 2007. Its ceilings have been dropped and its floors raised and its walls pulled out so many times to insert the wiring to connect it to the modern age, it is beginning to resemble floor 7 1/2 in *Being John Malkovich*. Strum squeezes between giant file cabinets in one of the warrens. His nest is stuffed with totemic objects and photographs, the giant poster from *The Seventh Seal*, with Death playing chess with Max von Sydow, and the print of the boy jumping over the puddle in the French train station by Cartier-Bresson, who died last year. There is a moody photo that Strum took of the expanse of the old Roman Senate—ruins and the pines of Rome. There are martini glasses and a pitcher on one of the book-laden shelves. The throbbing center of the room is a computer that with a few clicks calls up a screen dense with significant names: the consolidated files of every advance obituary for the *New York Times*, a choir of the not-yet-dead humming under their breath, beneath the obit editor's patient fingers. I may peek, I may see Mr. A— to Ms. Z— flickering past, but I may not mention A's full name, nor Z—'s. This is a matter of delicacy, the good manners of the premier obituarist for his future subjects, many of whom call to gossip about each other (*I saw him at a dinner party last night and he looked quite frail; Did you hear she was in the hospital?*) or to drop bits of info about themselves. The files are also secure from internal snoops, to keep other reporters and editors from filching details, hard won and written and edited to a "fare-thee-well," in one of Strum's phrases. There's fresh news in those files.

The notes from my sessions with Strum are pockmarked with "OTR"—off the record—every time he invokes a person who is not yet dead, in spite of the fact that some of these people have been interviewed for their obits. There is a sense of the sacred; these files are the vessels of the world's fragile but still living history. Most of those 1,200 or so names in there are of marked men and women, of course, consigned to this list by age or disease or high-risk occupations like the presidency of the United States. But it touches me to see them guarded so carefully, as if the obits were hearts that Strum will transplant to the obits page after their hosts are declared dead.

PORTRAITS OF GRIEF

Chuck Strum watched the collapse of the Twin Towers on television in the office on September 11, 2001, strategized quickly with the other *Times* editors, then hit the streets. He and a reporter had volunteered to check out the hospital emergency rooms. They started driving downtown from Forty-third Street, stopping at a hospital or two. Strum had a feeling there wouldn't be many wounded, and he was right. They drove as far south as Nineteenth Street, then parked and continued on foot. Mayor Rudy Giuliani's emergency bunker had been destroyed, but they found the makeshift command center and another *Times* reporter working there, covered in white soot.

The managing editor's "instinct, and not a bad one, is that we might be overwhelmed," said Strum, who at that point had been obituaries editor of the *Times* for a little over a year. "I knew that most of those who died in collapse would not have been news obituaries under ordinary circumstances. What I wanted to know, and could not immediately get an answer to, was what it was the paper wanted to do about all the others." The Metro section took the lead, appropriating for the moment the prominent 9/11 dead. In the days after the disaster, the *Times* ran a regular obituaries section in the back of the paper and another within the news coverage of the disaster. The obits in these two sections were similar in all ways except in the cause of death. Their subjects were the movers, shakers, leaders, brokers, and beautiful people who usually command obituary space, the figures with public lives. They contained all the usual parts and elements, and had the usual sparkle of a *Times* obit. But one section ended in the classified death notices, and the other appeared on a page with the Emergency Numbers to Call.

The oddness of these 9/11 obits was most pronounced in the phrases describing the cause of death. Obit writers from rival papers like to complain about the "clutter" in the *Times*'s obits, phrases like "according to his agent" or "as she said in an interview in this paper in 1987," but the *Times* lives and dies on its attributions. Someone has to announce every death, and then the writer of the obit has to confirm it. All quotes have to be nailed to the wall—where and when was it put on the record? "I want to know where you got this," Strum says reasonably. "You don't have to say 'the police said' after every paragraph, but you have to gracefully say how you know this." An obit, like any news story, is subject to the same rules of straight journalism the rest of the paper follows. One of the challenges of running obituaries for any of the 9/11 dead was establishing death; if it couldn't be a *verifiable* certainty, then it had to at least be a *veritable* one, and all available authorities were called upon to swear to it.

David Angell, a creator and executive producer of the NBC series "Frasier," died on Tuesday on American Airlines Flight 11 from Boston. He was 54 and lived in Pasadena, Calif.

Mr. Angell was returning with his wife, Lynn, 52, to California from their summer home in Chatham, Mass., NBC reported. Flight 11 was hijacked and flown into the north tower of the World Trade Center in Manhattan at about 8:45 A.M.

Kevin Scully, a family friend and funeral director, said the family had received confirmation of the death from the airline.

David Alger, whose approach to investing in stocks propelled the mutual funds he managed to the top of the 90's bull market, died in the collapse of the World Trade Center, his wife, Josephine, said yesterday. He was 57 and had homes in Manhattan and Tuxedo Park, N.Y.

According to Gregory Duch, the executive vice president and chief financial officer of Fred Alger Management Inc., none of

the 35 employees in the firm's 93rd-floor office in the north
tower at the time of the disaster appear to have escaped.

Berry Berenson Perkins, the widow of actor Anthony
Perkins, was in the first group of 9/11 obituaries that appeared in the *Times* two days after the attacks. Hers was one
of those delightful *Times* obits that weighs and measures the
life and also gives the reader a sense of the subject's milieu
and times. Berry, whose real name was Berinthia, and her
model-actress sister, Marisa, were related to designer Elsa
Schiaparelli and critic Bernard Berenson; they had been edu-
cated abroad. Diana Vreeland, who published Berry's photog-
raphy in *Vogue*, used to call the girls Berengaria and
Mauritania, after two Cunard ocean liners. Berry was a privi-
leged child who grew into a free spirit: she had married
Perkins in bare feet and a granny gown. The writer, Cathy
Horyn, had listed Perkins's sons, Osgood and Elvis, as survivors, then added an update on Berry's lifestyle: "In recent
years, Ms. Perkins spent time in Jamaica, where she ran a
beachfront bar with her boyfriend."

Flavored by the offbeat details of her pedigree and her life,
the obit seems modern and lively, but it is the standard *New
York Times* obit that has served the paper for dozens of years,
classic in form, with all its parts in place. "Berry Berenson
Perkins, a photographer and eclectic fashion plate of the
1970's before she settled into marriage with Anthony
Perkins," had died on "the first jetliner to strike the World
Trade Center, a spokeswoman for the family said." The *tombstone*, then *the bad news*, then *the song and dance*, with its taste
of her jet-set life as a *Vogue* photographer nicknamed for a

luxury liner. *The reverse shift*—"Berinthia Berenson was born in New York, part of a seemingly charmed European family"—and *the desperate chronology* were handled deftly, with details flavoring her wedding to Perkins; the only quote is not exactly a sizzling *friar*, but poignant under the circumstances: "I'm so delighted with my life," she had told the *Times* in 1977. "I have this fabulous husband, the man I always wanted to marry. I have two fabulous children, which I always wanted, and we're all so happy." The *lifeboat*, followed by a brief coda evoking the beach, ended the obituary, and the record of Berry Berenson Perkins's life and death was complete. There was no pathos in it, except that which the reader brought: the happy marriage ended with her nursing Perkins as he suffered his ultimately fatal AIDS-related illness in secret, and the charmed life ended in an explosion of jet fuel. In fact, two days after the stunning and consuming catastrophe, reading her obituary was like tripping into a fairy tale.

Her obit, and Alger's and Angell's and perhaps a dozen others, was surrounded by articles about the aftermath and the grief and sentiment that were part of the 9/11 story. The Metro section was struggling to put into words and numbers what it was we had lost: the densest zip code in the country, a significant percentage of the jobs and commerce in our largest city, our complacency. Years later, I still remember some of those stories. In one, the reporter spent an afternoon with the wife of the cofounder of a small investment banking firm. More than a third of her husband's employees had died in the attack, and she'd resolved to attend the funerals with or without him. There were sixty-seven of them. Another story described the Middletown, New Jersey, train station, where

190 cars were left after the last train pulled away on September 11th. These weren't stories about what the region had lost; they were about what *we* had lost. Even while it was running stories that observed the blurring of the lines between television journalism and the people who delivered it—reporters who had lost coworkers and family themselves, and survived or witnessed the attacks and now wore American flag pins and had tears in their eyes—the *Times* was subject to the same bias. Thousands of *us* had just been killed brutally *here*. FELLOW AMERICANS OPENING HEARTS, WALLETS AND VEINS read one early headline from the *Times*. Firemen were standing in the middle of the road, soliciting donations for the families of the firefighters who died—343 of them, the most devastating loss of life to an emergency response group in history—and it didn't matter if you'd given yesterday, the day before, or just a minute ago. You gave again.

Institutions like the *New York Times* gave, too. What to do about all the ordinary people who were missing and presumed dead? The *Times*'s response was ingenious. It didn't alter its obituary page, with its enduring but dispassionate structure, and it didn't bend the form to fit the circumstances; it created an alternative obituary form. "The *Times* became a local paper doing what a local paper should do," said Strum.

Its new section, A Nation Challenged, was designed to follow up on the attacks and cover news of the developing war in Afghanistan. In the back of this section ran Portraits of Grief, brief evocations of the victims that were at first called Among the Missing. As the *Times* explained above the first Portraits:

The official number of deaths in the destruction of the World Trade Center stood yesterday at 201, and the medical examiner has identified only a few dozen victims; almost 5,000 are listed as missing. Those glimpsed on this page include six who have been confirmed dead . . .

A Nation Challenged ran upside down on the back of the sports pages, which allowed the *Times* to give both subjects a front page. But it was an oddly resonant juxtaposition: you could read the sports news, once the teams started playing again, then turn that world of simple rivalries upside down to read stories about being embedded with the troops, or about the city's efforts to remove tons of shifting and still burning debris, or about the (mostly) young people who had suddenly disappeared one beautiful late-summer day.

In the chaos after the collapse, it had been almost impossible to determine who was dead. Images of "the missing" immediately began papering the island, hand-lettered posters or flyers made on home computers, in hopes that a legion of amnesiacs were wandering around in shock, unable to identify themselves, rather than lying beneath the rubble. Each day an updated and ever-changing count of the dead and missing appeared in the *Times*, but even a year later, there was still uncertainty in the numbers. (The 9/11 Commission ultimately settled on 2,973 deaths, the nation's largest loss of life on native soil "as a result of hostile attack in its history.") "Because it seemed impossible to write about the dead without confirmation, we decided simply to start writing about the missing," Janny Scott, a *Times* Metro reporter, wrote later. "And we decided to do it one by one."

Three days after the attacks, Scott and a handful of reporters had collected a stack of the fliers from bus shelters
and lampposts. Under the computer-copied faces on each
flyer were homely details, a nickname, the color of his polo
shirts and type of wedding band, birthmarks, surgical scars,
tattoos, which floor of which tower she had worked on, when
he'd last been seen or heard from—each followed by a number to call just in case. A photo and contact information—perfect starting places for the reporters. They began calling the
numbers, asking people to volunteer intimate details about
the missing, an awkward task at best. They listened to people
remember and weep, sometimes weeping along with them.
"It was heartbreaking work," Scott wrote.

This was the atmosphere in which Portraits of Grief was
created. Their language was casual, focused on daily routines
and mundane details. The man who brushed his daughter's
hair into a ponytail. "It wasn't really, like, good," his wife recalled, "but he did it." The woman who collected "useless
frog paraphernalia." The man who left a note for his family
that morning, "I fed the dogs but not the fish." Christine Kay,
the thirty-six-year-old *Times* editor who devised the form of
the Portraits, said they developed organically, shaped by the
constraints of space and the fact that reporters couldn't write
about the subjects as if they were dead. "One thing we made
sure we did," Kay said, "was to get some of the best reporters
in the building, which was hard because there were so many
already being used to cover this huge story. I knew it was all
in the execution." Good people, devoted mates and parents,
attentive mentors, angels, Little League coaches in the hundreds, workers who always had smiles on their faces, men who

shoveled sidewalks for old people, and women who were loyal friends. Their fate added tension to every detail, and colored it all with regret, though Kay said she worried at the time that readers would find them flip.

It did not take Lee S. Fehling's mother long to know that she had a character on her hands. "You know when the doctor slaps you on the back and the baby cries?" said his mother, Joan Bischoff. "Lee came out laughing."

"You know the sweetest thing my sister [Lisa L. Trerotola] did?" said her brother, Paul Spina. "She was planning a surprise party Oct. 6 for my brother-in-law's 40th birthday. Sent the invitations and everything. He has no idea. Would you put that in the newspaper? I don't have the heart to tell him."

Kay was an obits reader but said she wasn't influenced by the obits; she was inspired instead by a column that ran profiles of local figures called Public Lives in the Metro section of the *Times*. "Public Lives is also this crazy eclectic form, and what always struck me about them is that the most successful focused on a specific aspect of someone's life and ran with it," she said. "This was a form that needed to encompass dishwashers from Windows on the World, firefighters, stockbrokers, and successful executives. We needed to find a level playing field," she said. "Sometimes you couldn't tell. Sometimes it surprised you—you thought you were reading about a dishwasher and you were reading about a trader. That's what resonated, I think, whether readers realized it or not."

Even those who had been elevated in death to the obituary

page were sketched for inclusion in Portraits. David Angell's obituary had mentioned all his Emmys for *Frazier* and *Cheers*; his Portrait mentioned his love for his wife, his new home on Cape Cod, and his golf handicap—an impressive two. David Alger's, whose *Times* obit had cited his Harvard education and his investment theories, sketched him holding court with young traders over apple pie in a coffeeshop. Berry Berenson's quoted a friend, "If there was ever a person who could be called a living angel, I think Berry was"—a quote I guarantee you'd never read on the *Times*'s obits page. But these Portraits "were never intended to be obituaries, at least in the traditional sense of the word," Janny Scott wrote. As Christine Kay put it: "They were more like the anti-obituary."

The Portraits still resonate in the obituary world. Almost three years after they first appeared, they were the subject of heated discussions at the obit writers' conference. "So goddamn sunny!" one obit writer from a western paper complained. They weren't obits, they were vignettes. They didn't say where anyone was from. Did they elevate ordinary pleasures to sacred touchstones? Were they condescending? Were they biased? Andrew McKie pointed out that, as on his obituary page in the *Daily Telegraph*, they carried no bylines, then he thundered, "If it isn't bylined, I want Olympian detachment!"

Thomas Mallon wrote an essay for *The American Scholar* that also found fault: "Day after day, a dozen personalities were obliterated with the Grief team's pastels," he objected. "The *Times* has repopulated ground zero with the citizens of Pleasantville, and the 'newspaper of record' has been patting itself on the back for constructing the world's largest sympathy card."

Is the soft focus in the Portraits a failure, or the source of their power? Maybe it's both. The Portraits developed spontaneously, when American sentiment and journalistic objectivity met to mourn a mass tragedy in the *Times*'s backyard. They rose up out of the rubble.

At the conference, a chorus of voices followed the critical comments with affirmations. The Portraits dignified ordinary people; they helped heal wounded families and a wounded country; they made people cry. Also, the special section they appeared in, A Nation Challenged, won a Pulitzer Prize in Service for the *Times*. The Portraits had been reprinted in papers across the country, and subsequently created new markets for this sort of egalitarian obituary. Readers across the country locked on to them. For better or worse, they influenced obituaries. They became a model around the world for other newspapers trying to cover the deaths of ordinary men and women. And they were responsible, admitted one critic at the obits conference, for the jobs of a half-dozen obit writers in the room.

The evening of September 11, Strum took a subway uptown to Seventy-second and Broadway and stopped at a drugstore that was open late to buy some fresh underwear and toiletries. He was going to stay overnight at the apartment of James Barron, an old friend from work; Strum was worried that if he crossed the bridge to New Jersey to go home, he wouldn't be able to get back into the city. "I had a drink with James and his wife, Jane, and we looked out at their great view of lower Manhattan," he recalled. "Lights were on. At that distance from the Trade Center site, everything looked peaceful. It was surreal. I slept on a sofa in his guest room. The next day, I borrowed a shirt from James and returned to the office."

By then, the dead of the World Trade Center had become part of the urgent mobilization of the Metro section. The obits section dropped back to cover mostly famous people who were dying in mostly ordinary ways. Approximately 120 *Times* writers, most of them from other sections, ended up writing Portraits. "An enterprise like that required its own staff and direction," Strum reflected in an email. "The Portraits were not obituaries, per se, at least not as the *Times* defines them. They were memorial sketches, if you will. You'll find very little in the way of skepticism or analysis [in them]. Portraits was a fine idea. It was right for the moment. It captured the mood. It created a pitch-perfect insight into the human tragedy that became the heart of all the posturing, politicking, and national and international wrangling that emerged from the Trade Center wreckage and that continues today. It was a smart thing to do."

Who didn't feel useless right after the attack? The people who worked in national security were useless. Doctors and nurses and rescue workers were useless. Even the obituary editor was useless. What was needed was poetry. "[I]n times of crisis it's interesting that people don't turn to the novel or say, 'We should all go out to a movie,' or, 'Ballet would help us,'" Billy Collins had told the *New York Times*. "It's always poetry. What we want to hear is a human voice speaking directly in our ear." That was provided at the *Times* by the reporters in the Metro department and published as Portraits of Grief.

Strum spent the night of September 12 at a midtown hotel where the *Times* had set aside some rooms for staff members.

"I tried to buy another shirt at Brooks Brothers, but it was closed. I ended up, I think, at the Men's Wearhouse, on or near Madison—don't recall. I bought a blue-and-white tatter-sall shirt with a straight-point collar. I still wear it and I think of that day every time I put it on."

GOODBYE!

Stephen Miller was working as a computer technologist for a Japanese banking firm on the eightieth floor of the South Tower of the World Trade Center the morning of September 11. Miller would be the first to tell you he was no computer expert—he had been a religion major at Oberlin College—but he knew more than the other people at his firm, and after six years on the job he was comfortable. The job paid well, but it was, you know, *boring*, and Miller was one of those shaggy guys who tend to settle on the fringe of academia, a restless intelligence in search of something to obsess over or write about—the weirder and more offbeat, the better. He had grown up next to a cemetery, a great place to ride your bike and get stoned, and after college he spent some time plugging names and job histories into formula obits on the night desk of more than one little New Jersey paper (bored, he took to filing fake obits: "I killed friends of mine," he joked). With the exception of a few glints of life and humor in McG.'s obits in the *Times*, obits, he felt, were a neglected, hackneyed form, ripe for subversion. So Miller helped bankers edit their desktops all day, then went home to research and write darkly amusing alternative obituaries. For half a dozen years, he juggled his straight job and his twisted hobby. Periodically, maybe four times a year, he sent his friends and a growing list of fans a batch of obits in the form of a 'zine, a glorified newsletter he called *GoodBye! The Journal of Contemporary Obituaries*.

Closer to essays than classic obituaries, *GoodBye!*'s sendoffs left out some of the conventional features of obituaries. You wouldn't find lists of survivors in them, for instance, and they often failed to cite the age of the deceased. "So what?" as Miller would say. He wrote about oddballs like an expert on Bigfoot, or the carny sideshow act Melvin Burkhart, "A Freak with a Nail Up His Nose." He looked for the people lurking behind the news, like Jeffrey Dahmer's mother:

What is every mother's nightmare? Is it squashing junior while on a bender? Is it driving off a cliff with junior in the back? Is it sending junior off to an insufficiently researched kindergarten that turns out to be a front for NAMBLA? It is of course any of these, but they sound like teething pain compared to a nightmare few mothers could have had: junior turns out to be a murderous homosexual cannibal. And then come all the inconvenient, impolite questions, because everybody wants to know why.

He named the elephant in the room, and he riffed on it. In his obit of sex researcher Dr. Alex Comfort, Miller elaborately compared *The Joy of Sex* and *The Joy of Cooking*. He regularly ran a roundup of people who had died doing something stupid—the so-called Darwinian deaths. Where the *New York Times* would use a light brushstroke to liven up its obits with a funny detail, or throw an exotic mule diver or lighthouse keeper in with the lawyers and judges, *GoodBye!* hammered the nail up your nose. A decade after the London obit had developed into a form of entertainment, Steve Miller was cooking up his own version of the opinionated, storytelling obit. When a friend brought him the first collection of *Daily Telegraph* obituaries, he was knocked out; *this* was what he was talking about. He began describing his work as belonging in the British tradition. "*GoodBye!* followed the British model of laughing at the proud and including as much scandalous detail as possible," he wrote. By the fall of 2001, *GoodBye!* had a few hundred paid subscribers, a steady stream of hits on its website, and a profile in the world of obits freaks.

At the age of thirty-nine, Miller was on joking terms with death. Little skulls decorated the pages of his 'zine. He had a new wife, a pretty paralegal. That morning, September 11, he was wearing a new pair of shoes that he hadn't yet broken in. In his pocket was two hundred dollars, the proceeds of a "dead pool" he had organized with some of the bankers in his office; they had been gambling on which celebrities and newsmakers would die next. Were there any other ironies he could stuff in his pockets and take down eighty flights of crowded, smoky stairs? From his recollections of that day, delivered as a speech to other obits writers and posted on his website www.goodbyemag.com:

> *I remember the squealing sounds the building made while I walked down the interminable flights of stairs, the heat that made me sweat through my clothes and the eerie yellow light that suffused the stairwells. I remember feeling dizzy then,*

wondering if the vertigo was caused by the building falling or just my own nervousness. In the stairwells we really had no idea what had happened—no idea that airliners had slammed into the towers. All we knew was that it was something bad. . . .

When the stairs got too crowded, he stepped onto one of the floors to look for a phone to call home. That's when he saw the north tower with its gaping, flaming hole and bodies tumbling past. "Oh my God, they're jumping!" the people at the window were screaming. He couldn't reach his wife. He found a less crowded staircase to descend, made it out of the building just before it came down, and scrambled his way through the havoc without being hit by anything. He walked across the Brooklyn Bridge to a neighborhood where garbage-men were "still collecting the trash while dust from the collapse of my office rained on them, as if they were working in a snowstorm." Incredibly, he ran into his wife on the street. How anyone survives such a day, while friends and coworkers are dying around him, is a story, and the story of this sardonic wise guy, the trafficker in obits, is no more or less incredible than anyone's. The ironic forces fell short of killing him.

"Writing a large number of obits does not in any way qualify you to deal with death on the scale of wholesale carnage," he wrote with the spooked humility of a survivor. He told his story to various members of the press, to the Fourth Great Obituary Writers' International Conference the next spring, and to fans of his website. It troubled him that he couldn't keep the chronology of the day straight or precisely describe the screams of the witnesses who suddenly realized the other tower was boiling over with people. The day had been so vivid, and his memories were so fractured. He wondered what that meant about his, or anyone's, ability to represent a person's life in several dozen inches of text. "If a few minutes of my own life, moments of irredeemable clarity that spanned at most a couple of hours, are so difficult to get right, how much harder is it to present a truly accurate version of an entire life in 20 or 30 newspaper inches?" he wrote. And what sense did narrative make when one's life story could be interrupted so arbitrarily?

But instead of putting him off obituaries, the attacks of 9/11 seemed to focus him. He sent old issues of *GoodBye!* to the *New York Times* and a few other newspapers, offering his services as a writer. The *Times* passed with the excuse that they didn't hire from the outside. But his efforts paid off a little more than a year after the attacks. As he was hurrying out the door with his wife—her water had just broken; they were headed to the hospital for the birth of their son—the editor of the *New York Sun* called Miller to discuss a job writing and editing their obits. A systems job on Wall Street that would have paid twice as much materialized at the same time but didn't really tempt him. His hobby had become his calling, maybe even his destiny.

ATTENTION MUST BE PAID

Three and a half years after he survived 9/11, Stephen Miller, now the obituary editor of the *New York Sun*, finds me in line outside the Majestic Theater, a block from the *Times* building in the middle of Manhattan, waiting for the doors to open for Arthur Miller's memorial service. Steve (no relation to

Arthur) is nearly forty-three. He wears cool shades, khakis, a pair of bud earphones in his ears, a messenger bag slung over his shoulder. He has dimples and flashes them frequently. He didn't have a cell phone on 9/11, and he doesn't carry one now.

He has curbed his humor and suppressed his more mischievous impulses to work for the *Sun*, a weekday newspaper founded by two conservative editors from the *Jewish Daily Forward*. He's gone legit, or anyway, mainstream. He writes substantive obits of Johnny Carson and Artie Shaw, does the occasional tribute to a conservative billionaire, and has his fun with the oddballs, the rare-books dealer who discovered that Louisa May Alcott wrote stories about hashish and murder before she settled down to *Little Women*, the B-movie star who retired to the B-life:

> *Although it turns out she was hardly the South American bombshell she was publicized to be, Acquanetta went on to live in the shadow of her years of demistardom, a kitschy celebrity in her new hometown of Scottsdale, Ariz., where she appeared for years on local television ads for a car dealer and raised money for local charities. She said her name meant "burning fire and deep water."*

He doesn't try to cover everyone of importance who dies. Instead, he focuses on one obit a day, and fills in with obits he picks up from the *Daily Telegraph*, the *Los Angeles Times*, the *Washington Post*, or the wires, though he misses altogether people who, like Arthur Miller, die too late for Friday's paper and too early for Monday's. A story about Miller's memorial service will be a nice way to catch up.

Is there any better entertainment in New York than a celebrity's memorial service? It's no secret; the public is welcome. A notice appears on the obits page of the *Times* telling you when the doors will open and who has promised to speak or perform, to lower the curtain on a public life and stage a personal tribute: people you've heard of, telling stories you've never heard. Like obituaries, memorial services are a garden of emotion, with laughs cropping up behind the tears. It is the ultimate theatrical experience, united with your fellows in appreciation of someone unforgettable.

And that's not to mention the production values. Here in this city, the sound quality is excellent, the pianist performs professionally in the theater. The photographs projected on the giant screen are first-rate. The short montage of film clips rivals those shown at the Academy Awards. Every so often, someone like Estelle Parsons takes the stage to give a dramatic reading. The seats? Well, they're on Broadway, plush velvet, banked for quality sightlines. And—awesome detail— the whole experience is free.

It is a gorgeous day in May. In the long line for public mourners, people are sunning their faces, grazing through newspapers, chatting each other up, but the line of invited guests enters first; by the time we climb the opulent staircase inside and find seats, the orchestra is half-filled with Arthur Miller's friends, relatives, and colleagues. Like the other voyeurs, we settle in the back, on the aisle, and get up half a dozen times for others as the theater fills—aging men in suits or jeans, women with tousled gray hair in sweatpants and sneakers. We have a few minutes to study the program, studded with famous names. "*Bill* Coffin?" Miller says with a laugh. "William Sloane Coffin? He's almost dead himself." When the lights go down, everything but the stage becomes very dark. Only the emergency exit lights on the aisle floors are lit. Miller meant to bring a flashlight but forgot, and we begin the comedy of trying to identify the parade of speakers and performers who appear under the huge black-and-white projected image of the deeply etched face of Arthur Miller, "a giant of the theater" and "one of the great American play-wrights," as Marilyn Berger called him in the *New York Times*.

"Who's that?" hisses Miller, referring to a man who looks like either an accountant or an older actor who plays one. "Don't know," I hiss back. The accountant-man is saying, "I read Arthur Miller's obituaries. Just about everybody got it right, except the *New Criterion*, which was so happy he had died," and the lefty crowd mutters its disapproval. Miller and I have both been taking notes in the dark. He leans into the aisle and steadies the program under the exit light. "Albee!" he says in amazement. By this time the playwright Edward Albee is leaving the stage after some closing inanity: "Some playwrights matter. Arthur mattered—a lot." We have to forgive him his relative lack of eloquence. At the bottom of the *Times*'s obit page, next to the announcement of the Miller memorial, was a brief obituary of Jonathan Thomas, a sculptor who died of bladder cancer, according to "his partner, the playwright Edward Albee." Albee's lover had died just a week ago.

We both recognize Daniel Day-Lewis, Miller's son-in-law, who reads a funny story Miller wrote about delivering bagels and onion rolls by bicycle, hazardous work in winter. But who is the intense man who takes the stage and speaks rapidly about watching his mother playing Linda Loman in a community production of *Death of a Salesman*? Even though he was only six at the time, the man says Miller's words made "my heart break and burst into flame." Soon we realize he's Tony Kushner, another great political playwright, who won a Tony and a Pulitzer for his pair of plays *Angels in America*. Kushner tells us he sat behind Miller during one Tony Awards ceremony and spent the entire evening gazing in wonder at the domed head in front of him, the source of all that great American drama. "I wanted to touch the head, but thought the owner might object." The theater ripples with laughter.

There's a professional gloss on the contents of the service, what you'd expect for a public figure, affectionate glimpses of Miller on walks or laughing and arguing at the dinner table, but no danger of falling into anything really personal or mawkish. Miller's son doesn't rhapsodize tearfully about a Little League coach who taught him to ride a bike; instead he reads his father's defiant letter to the House Un-American Activities Committee in a steady voice. And Miller's occasionally messy personal life has been edited: his third wife, Inge Morath, who died in 2002, is invoked frequently, but his first wife, mother of two of his children, is barely mentioned, and no one breathes a word about Marilyn Monroe, to whom he was married for almost five years, or Agnes Barley, the abstract painter fifty-five years his junior who lived with him at the end.

I realize spending a beautiful afternoon listening to this orchestrated fugue of sorrow and laughter for a departed literary figure is not everyone's idea of entertainment. But on May 9, there's no better place to read the complex layers of our culture than this memorial service on Broadway. Between odes to Miller the brilliant dramatist and Miller the stalwart leftist, speaker after speaker invokes the great line from *Death of a Salesman*—"Attention must be paid." Nobody was more of a champion of the common man than Miller, who grew up in Brooklyn the son of a successful man ruined by the Depression. In fact, Miller's account of the death of Willy Loman can be read as an elevated obit of a common man. George McGovern, who credits Miller with helping him win the Democratic nomination in the 1972 presidential election, saw a production when he was still in school, he says from the stage; he recalls an audience of grown men with heaving shoulders trying to stifle their sobs. The American "everyman" haunts us all. What a tragedy to slip into death without fanfare or tribute! Miller himself was quoted in an article published in the *New York Times* a few weeks after the play's Broadway debut: "I believe that the common man is as apt a subject for tragedy in its highest sense as kings were." He described Willy Loman as a man who "was trying to write his name in ice on a summer day."

The lights go up. We walk past the posters and paraphernalia for *The Phantom of the Opera* and into the sunshine, find a Ranch 1 on Eighth Avenue, and continue a conversation we started almost a year ago at the obits conference. We tend to circle back to the same topics, particularly the common-man obits, which have become increasingly popular around the country since 9/11. Willie Loman crops up now in Portland, San Jose, Denver, Cleveland, Phoenix, Philadelphia, and Washington, D.C., some well written, some full of clichés and naked appeals to sentiment. Steve Miller dismissed the work of one emotional obits writer as "lurid prose and ordinary people. I recognize there is a place for this sort of thing, that there's a market for it, that many people I have respect for believe in this. But it ain't me, babe."

Miller's own obits for the *Sun* attempt to cross the reportage of the *New York Times* with the *Daily Telegraph*'s wit and spin. He's borrowed the *Independent*'s trick of putting the grind of the vital statistics in a separate box, and the London papers' emphasis on striking photos to accompany the text. He's always good for a cheesecake shot, babes from the forties, poses that are arty or funky—anything striking that doesn't look like a high school yearbook photo—and this alone sets him apart from the *Times* with its old-fashioned layout and preponderance of head shots. Since the obits conference, he and Adam Bernstein, the Kid from the *Washington Post*, have been sharing sources and keeping up a running dialogue of wisecracks and breaking news ("It's great. We both compete with the *New York Times*, but not with each other"). Amelia Rosner, the alt.obituaries maven, frequently alerts Miller to overlooked subjects. His inspirations are sophisticated and his colleagues run with the sardonic crowd.

Miller is, as far as I can tell, allergic to sentiment. I've never seen him get sappy in print, not even when he was writing about surviving the burning towers. But though he says, "I like to write what I like to read, and frankly, I don't like to read obits of average people," he admires those average-people obits that are well done. "I don't really recognize them as obits, but they're awfully interesting. I think we're all interested in the fact that there are different visions." As for the emotional Portraits of Grief: "In some ways I thought they were silly or frivolous. I felt like they were too one-dimensional, too cheerful, too political," but with several years' perspective, he thinks they've aged well. "They seem like a semi-artistic act of mourning—not obituaries at all, but an attempt to put flowers on the graves."

Though we are a block or so from falling in its literal shadow, the *New York Times* shadows all our conversations, the way New York City shadows Philadelphia. The small-circulation paper and the smaller city are obsessed with their mighty rivals, though Philadelphia barely exists to New Yorkers, and the *Times* ignores the *Sun*. ("Do you read the *Sun*?" I asked Strum. "I don't have time," he said curtly.) Miller, however, is a mosquito who enjoys nibbling the *Times*.

After our chicken Caesar salads, we linger by the subway he'll take downtown to his office. I could stand on the street corner discussing the relative usefulness of various news data banks with him for hours (he's a Proquest fan; I like Custom Newspapers). I try to imagine the contemporary history of obits if he hadn't escaped death, if he were two hundred words in a Portrait instead of a contrary presence on the New York scene. It would have been hard to capture him in a Portrait, the tech guy who went home at night and wrote funny obits, and a mockery if he'd ended up the subject of our pity.

I linger at a newsstand before taking the train home. New York is the home of numerous newspapers: the *Daily News*, the *New York Post*, the *Wall Street Journal*, the *Observer*, *Newsday*. The *Post* takes potshots at the *Times*'s obits, and gleefully reports its errors (unearthed originally, one suspects, in the *Times*'s own corrections columns); when the *Post* runs the occasional obit of someone on staff, it reads like a note posted in the coffee room. The *Daily News* no longer runs obits. After the terrorist attacks, the *Journal's* offices had been evacuated, yet its staff still managed to put out an award-winning paper on September 12 from a plant in New Jersey and a New York City apartment. The *Wall Street Journal* could do anything it wanted to, including putting out a first-rate obituary page. But it doesn't.

Besides the *New York Times*, only *Newsday* runs interesting, full-blown obituaries. It reprints stories from wire services and other sources to cover the notable deaths, and each day it features a staff-written obituary of someone from Long Island that assesses the subject professionally and gives an intimate glimpse of his life. From JAMES JOSEPH FINNERTY, 81, SURGEON, PILOT, FISHERMAN:

> *A lover of words and literature, Finnerty often quoted sayings in French, Latin and Swedish, his son said, and never missed The New York Times or Newsday crosswords. He also loved to travel and cook, and "he made the best damn gravy," Charles Finnerty said.*
>
> (*by Indrani Sen*, Newsday)

Two days after 9/11—days before the *New York Times* began its series—*Newsday* was running its own biographical portraits, called The Lost, "of people missing or presumed or confirmed dead in the terrorist attacks in New York, Washington, D.C., and Pennsylvania." The short pieces sketched the individuals, put their presence on the scene of the terrorist attacks in context, and tried to impart the flavor of their lives: the young man who had run with the bulls in Spain, the couple on their way to Hawaii. One bio quoted the man's brother: ". . . there is still hope he's under the rubble, breathing." *Newsday*'s collection of The Lost, with additional stories by other Tribune newspapers, is called *American Lives: The Stories of the Men and Women Lost on September 11.* It's harder to find than *Portraits: 9/11/01*, and its stories have been edited and updated since they first appeared, but they have a familiar, heart-tugging ring.

> *[Brian] O'Flaherty likes to think that his friend [Dennis Cross] is in heaven, talking about firefighting with Cross' father, who died of a heart attack after battling a major fire when Cross was 13. "He probably told his father all the new tactics they now use in the department," O'Flaherty said. "They probably laugh at it all, because nothing has changed. The fireman still crawls into the building to put out the fire the same way as 50 years ago."*

> *[Karen Klitzman's] family is contributing money to the fellowship fund and is asking friends and other family members to donate money as well. "Something more worthwhile has to come out of this horrible tragedy," Donna [Klitzman, Karen's twin] said.*

Newsday was overshadowed by the *Times* in the aftermath of the disaster, and also when the awards were handed out. The *Times* swept up seven Pulitzer Prizes for 2001, including one for Public Service for its special section, A Nation Challenged, which included Portraits of Grief. *Newsday* was recognized by the Pulitzer Prize committee, but not for its 9/11 coverage and its improvised obituaries—for its classical music coverage.

5

Now You Know

You don't know how many things you don't know until you start dawdling over the obituary page. From tiny pieces of cultural flotsam to profound illuminations of history, facts and factoids spill out of the columns of the departed. What subject will divert us today, the Bataan death march or the discography of Beatles? One day you're reading about a dissipated celebrity, the next day you're getting a lesson from "Verne Meisner, 66, Musician Who Championed the Polka." Meisner, pictured grinning over his accordion, had been responsible for dozens of albums and inducted into no less than five polka halls of fame. Who knew there was even one?

The demand for polka is insatiable. . . . Today there are polka parties and polka clubs, polka conventions and polka cruises, polka chat rooms and polka tours, polka museums and polka memorabilia. Concentrated in the Upper Midwest, the polka subculture spans the country, with tens of thousands of adherents, perhaps more.

(by *Margalit Fox*, New York Times)

Reading this took my breath away. Polka, sweeping the world! I lived in the upper Midwest as a teenager, and had no idea; somehow I escaped the cheerful stomping and hooting of a crowd possessed by this particular virus. I never saw the roof of a VFW hall jump with the joy of polka. Where had I been? Never mind. I know now. And I have an inkling of how large and varied are the tides that crash through our culture.

I can no longer pretend ignorance about roller derby. The death of Ken Monte, seventy-five, was an occasion to learn about that which we were assured was "every bit as much a legitimate sport as pro wrestling." Exactly!

The derby had been invented in the 1930s by a Chicago promoter eager to capitalize on America's infatuation with roller skates during the Depression. In the derby, teams of five skaters circle a banked wooden track and score points by lapping and forcing their way past one another. . . .

A terror on the banked track, Mr. Monte was known for his speed and skill as much as his flying elbows and crushing knees.

(*by Steve Rubenstein*, San Francisco Chronicle)

I prize these glimpses of another world. From "Buck Johnson, 85; a Maker of Deer Scent," by Trudi Hahn of the *Star Tribune*, I learned that young Buck used to listen to his grandfather "tell of Indian hunters who used the musk glands from animals to cover their human odor." So Buck experimented until he hit on a great formula using secretions from a gland between the toes of a deer.

So now I know how to catch a buck. I know how to catch other creatures, as well. In Margalit Fox's appreciation of James

A. Houston, an artist and author of *Confessions of an Igloo Dweller* and other books, we learn how to lure a wolverine—with a Kleenex soaked in perfume.

Some call it trivia.

Elton John used to be named Reg Dwight, but changed it in tribute to the British blues-rock singer John Baldry, we learned when Baldry died.

"Johnny B Goode" was written by Chuck Berry to salute his overshadowed, squeezed-out partner on all those early rock-and-roll records, the great pianist Johnnie Johnson, who died at eighty.

Carly Simon probably wrote "You're So Vain" not about James Taylor or Warren Beatty or Mick Jagger, but about the dissipated eccentric William Donaldson, who left her "when she was still quite naive." Donaldson wrote wonderful satirical books, but he also ran through several fortunes, pimped, and enjoyed crack cocaine and the date-rape drug Rohypnol (he liked to use it on himself). "It's such a nuisance," the *Daily Telegraph* quoted him in his obit. "The trouble is, it wipes your memory. You have to video yourself to appreciate just what a good time you had."

Who knew?

Stories of World War II, rock-and-roll, Vietnam, civil rights, and the women's movement are being added to and filled in every day. Tracking these threads is like watching our history as it's woven into pattern. The last of the Comanche code talkers, Charles Chibitty, died at eighty-three. Perhaps you knew the Navajos used to send messages for the Allies in the Pacific during World War II. Comanches did the same for the Allies in Europe.

You never know when a shard or two from our messy history will turn up. In the *New York Times*'s farewell to a coin dealer, Douglas Martin listed some of the other items John J. Ford, Jr., collected. These included the medals that presidents gave Indian chiefs, no longer worthless, and "the badges slaves wore when they were rented out for day work." There's a piece of our hidden past, strange as the bone of an extinct animal.

The obit pages were cheated of the biggest story from Watergate. The identity of Deep Throat, the mysterious source who contributed so much to Bob Woodward and Carl Bernstein's investigations into the scandal in the *Washington Post*, was supposed to be revealed upon his death. This tantalizing promise came from Woodward, who had written Deep Throat's obit in advance (both as a newspaper story and at book length) and kept it under lock and key for the big day. Then W. Mark Felt, a former deputy FBI director, jumped the gun. He was ninety-one and suffering from Alzheimer's but remembered sneaking into the underground parking garage to whisper government secrets. Felt's motives as Deep Throat were both heroic and petty; it seemed he had wanted to be FBI director but Nixon appointed L. Patrick Gray III instead.

Gray died a month after Felt's revelation, and the former director's obit made much of Gray's bitterness, toward both the president who lied to him and the deputy who betrayed him. Gray had left public service under a cloud, liquidating his personal assets for his legal defense. He had resumed practicing law, Todd S. Purdum wrote in the *New York Times*, but not as L. Patrick Gray III—as Louis P. Gray.

The river of history that flows through the obituaries is full of ugly currents. A critical-care doctor and writer in Chicago, Cory Franklin, collects quotes from obituaries, and while combing the papers in July 2002 uncovered a tragic vein. Rosemary Clooney, the beloved chanteuse, had died at seventy-four. Her wonderful career had come apart in the late sixties, with a bitter divorce from actor José Ferrer, and the assassination of Robert F. Kennedy. She had supported RFK's bid for president; in fact, she had been standing next to him when he was shot. Afterward, she spent a month on a psychiatric ward, and, according to Richard Severo in the *Times*, "when she emerged, she was short of money and supported herself by singing for anyone who would pay her. Mostly it was weekend work at Holiday Inns." She made a comeback, eventually, but the days of the mega-hits were over. RFK's death derailed her.

A week after Clooney, the director John Frankenheimer died, too. By the late sixties, the director of *The Manchurian Candidate* and *The Birdman of Alcatraz* was considered one of the era's most important auteurs. Frankenheimer, who had been responsible for RFK television spots, drove the candidate to the hotel where he was shot, and this terrible experience, according to Brian Baxter in the *Guardian*, "initiated the director's deep depression." Frankenheimer recovered enough to direct *The French Connection II* and other films, good and not so good, but, like Clooney, he and his career had been devastated.

I thought of this dark parallel when, three years later, I read Fred Dutton's sendoff. Dutton had been an aide to California governor Pat Brown before joining John F. Kennedy's administration. He continued working for the Democrats and the Kennedys after JFK's assassination.

When Robert F. Kennedy ran for president in 1968, Dutton
served as behind-the-scenes campaign manager. He was with
the candidate when he was slain at Los Angeles' Ambassador
Hotel, and rode with Kennedy and his wife, Ethel, in the am-
bulance to the hospital. . . . "After Bobby was shot, the lights
went out for me," he told The Times in 1981.

                    (*by Myrna Oliver*, Los Angeles Times)

The caption on the *Times* photo also made the point deliberately: "Shaken by Robert F. Kennedy's assassination, he
eventually became a lobbyist so he could stay involved in politics at a less personal level." Dutton went on to private law
practice, and became so effective negotiating arms sales for
the Saudi Arabian government that he earned the nickname
"Fred of Arabia." This would have been a baffling career
move to those who hadn't followed history as it plays out in
the obituaries; but for those who had, that profound turn was
familiar. The Ambassador Hotel had been the scene of one
sudden death—and many slower deaths the obit page is still
catching up with.

6

===

Ordinary Joe

I was sitting in my comfortable bed outside New York City, reading a story cobbled together by three staff writers about a beheaded screenwriter in the *Los Angeles Times*. The ninety-one-year-old victim had written *Abbott and Costello Meet Frankenstein*, then had been blacklisted. He moved to Tucson and became a maître d', then returned to L.A. to write TV shows pseudonymously, leading an active life until, obviously, the beheading. The killer had climbed a fence with the head under his arm and then broken into a neighboring house, and stabbed a doctor who was on the phone making reservations for a trip to San Jose. I rarely read news stories to the end, but because of its mix of incredible and prosaic details, I couldn't put this particular one down, and I couldn't resist sharing it. Listen, I told my husband: "[The perpetrator] was apprehended as he sat on a wall under a row of ficus trees near Melrose Avenue. He had a Bible and a small can of Mace. . . . 'He seemed like a perfectly normal guy,' his landlord said, 'but he was always in a rush.'"

And there were more details. The screenwriter's body had been found under blankets in his bedroom by his eighty-six-year-old girlfriend. "I was befuddled for a minute," she said. "It was like a movie, not real life."

I wouldn't have been surprised if hundreds of people were discussing that story just like us, horrified, but in a pleasurable way, savoring every one of the details. They make the story—the quotes that slant sideways, the homey, almost funny specifics, the deadpan delivery, the ficus trees shading the be-header as the police made their capture. They have a kind of Elmore Leonard glint to them. They're particularly L.A., and they resonate in that authentic way that most newspapers seem to capture more by accident than design. It's almost impossible to teach that sort of writing except by pointing students to a stack of clips and telling them, "Inhale these."

I think they come from the obituaries. I think they come, specifically, from the obituaries that started appearing in the *Philadelphia Daily News* in October 1982, when these bright shards of detail and glimmering quotes began to appear, attached naturally and unapologetically to the obits of regular people. People whose lives had been considered dull as linoleum to the general public were offered up as heroes of their neighborhood and characters of consequence. Even more important, every particular of their quirks and foibles— the brand name of their cigarettes, their taste in horror movies—was presented as a clue to the mystery of their existence in the fascinating story of their lives. *There was this guy in Philadelphia*, people kept telling me. *You've got to find him. Is he still alive?* His name was Jim Nicholson. It's hard to imagine Portraits of Grief, the common-man obits, and the columns remembering local lives that appear across North America now without him.

Nicholson plucked people out of the sea of agate type and wrote full-blown feature-style obituaries about them: a janitor, a grandma known for her love of poker, "a world-class scammer." There are lots of little newspapers in the United States where ordinary citizens are written up when they die, for readers who knew them. Nicholson and the *Daily News* gave them a big shot's space in a big city paper, without dressing them up to do it. The subjects were characters from the urban landscape, with nods to Damon Runyon and the hard-boiled city columnists, but they were also just regular folks, being written about in a natural way. "What I try to do—" Nicholson said in an early interview. "If two guys meet at Broad and Snyder and one guy says, 'Did you hear that Joe died?' And for the next two minutes, they'll talk about Joe and write his obit—he was a good pool player, had an eye for women, never broke his word . . ."

He figured out a way to make the obit porous and let some of the real world leach into the strict borders of the form. He was willing to explain in the course of announcing the death of an old neighborhood woman that "jitterbugs," or "bugs," were what people called the drug users who hung out on the corner. If it were a plumber's obit, he'd try to work in a practical tip, like the way to clear a clogged toilet is with hot water and Tide. He had the ability to both inhabit the world he was writing about and give it perspective.

They were married three months later and not because they had to.

It was one hot day when she watched her oldest son Thad plow-
ing that she decided her children wouldn't spend their lives fol-
lowing a mule through the South Carolina dirt.

In a lot of scenes in past years, she was the woman holding up
the scenery.

Society today does not assign extraordinary attributes to a 35-
year-old heavy-equipment mechanic who is living with his
parents and whose possessions do not appear to much exceed a
Miller Light and a pack of Marlboros on the bar before him, a
union card in his pocket and a friend on either side.

Between news of the death and the list of survivors,
Nicholson would slip in sly comments like, "He had the digestive juices of a shark," or "Charlie did it all with one eye."

And he elicited the most incredible quotes. "'I had unfortunately burned up my cat Smokey in the dryer,'" one story
began. "Lou gave me a book, *1001 Uses for a Dead Cat*. You
loved him and at the same time you wanted to strangle him."
Or, in the obit of a theatrical producer, who had given his
friend opening-night tickets: "It was the worst show I've ever
seen in my life. Country music and ugly women. I didn't leave
because Stan would have been upset. That's my greatest tribute to him. . . . Stan would understand."

One of his most quoted obituaries is of an unemployed
drifter named Thomas "Moose Neck" Robinson whose body
lay unclaimed for several days but who, it turned out, had numerous friends and family who had loved him on his own
terms. His niece remembered his "good heart. A lot of people

just couldn't understand him, what was wrong. But he heard and saw things we don't know." Moose's brother said, "He was interested in going around asking people, 'Have you got a dollar?' I'm not going to tell you a lie. Moose was a drinker. He'd go around and ask people for money, and they'd give him anything he wanted. Everybody fell in love with him."

When it came to sending off the practitioners of the tougher professions, Nicholson preferred the direct quote. The son of a veteran cop recalled, "It would take him two minutes to tell if a guy was dirty or not." He quoted one city editor about another: "Will could be real gruff at times. . . . You'd ask him a question and he'd either yell at you or mumble, 'I don't f——ing know.'" Other reporters recorded lines like that, but they didn't usually use them, and certainly not in the obituaries, a place reserved for the beloved and the devoted. Nicholson used them to advantage, to get past the polite veneer that usually glosses tributes to the dead, and to say things he couldn't say. His obituary of a handsome mail carrier evoked a ladies' man with quotes from the man's daughter—"My dad grieved hard when women would die and people wondered why"—and his first wife—"There were no flies on him, no place."

The best quotes were so much more than pithy lines; they were windows into a culture. From the brother of RICHARD "BOSS HOG" HODGES, SCHOOL CUSTODIAN, BON VIVANT: "'Cook?' said an incredulous William Hodges. 'His roast beef melted in your mouth. And fish and grits. His biscuits and cornbread talked.'" In Nicholson's hands, the dead shimmered with life. You could taste their cornbread.

My favorites were obits where Nicholson, obviously a

rebel, refused to write the expected. Adolph J. "Ade" Yeske, a man who could have been "a gentleman farmer in Bucks County," instead answered the Lord's call to minister to the poor in the Brotherhood Mission in a hard-luck section of Philadelphia—and was broken by the work. "Thirty-three years of serving soup, running the secondhand shop, preaching sermons to a sea of beard stubble, bleary eyes, drawn women and crying children would grind and burn him out physically and mentally." The silver lining that readers look for in hard-luck lives, and that Nicholson frequently uncovered, was not to be found here. Instead of manufacturing one, the writer became Ade Yeske's recording angel. "A fundamentalist, he preached right out of the Bible, and his nephew said he 'was not especially inspiring.'" He lost his patience when people he'd help kick the bottle would show up with liquor on their breath, and the ordinarily patient Yeske "could shove that person out the door in frustration." Yeske, whose vacations consisted of "day trips in an old Ford or Chevy he drove until it fell apart," eventually lost his health, and had nothing set aside for retirement. A pension was begged and scraped together, and Yeske spent the rest of his life passing out religious tracts at hospitals and nursing homes in Florida. "Curt Yeske said it rankled him that a man like his uncle was allowed to slip away without any real recognition or send-off. But by all accounts, Ade Yeske didn't mind because he never actually worked for them in the first place."

Nicholson must have tunneled into that sorry, Job-like world and seen what Ade saw, that being used up in the service of his mission *was* his mission. He wrote this one without sentiment, and then published the obituary three days before

Christmas. Yeske's story is all the more heartbreaking because the writer threw away his violin. Imagine this obit in the hands of a local newscaster, or one of the sappier chroniclers of ordinary people during Christmas week. It would have been milked for every oily tear. Instead, it reads like a piece of Steinbeck's bleak America, like Truman Capote or Joan Didion or any of the dozens of great New Journalists who had been bringing the texture of fiction, its telling details and vivid characters, to nonfiction since the mid-1960s. This guy understood the people he was writing about from the inside out and, somehow, made it clear that he was writing about people just like him. In nineteen years he found something extraordinary to say about more than twenty thousand ordinary Philadelphians.

And if all that didn't make him a legend, it seemed Nicholson had juggled his obit writing with a career in counterintelligence. "His spy stuff," one of his coworkers called it. First a captain, then a lieutenant colonel in the army reserve, he'd take a few months' sabbatical every year to go away on clandestine assignments in Panama or Tajikistan or on the Mexican border. He'd sit at his desk, interviewing grieving people and writing up obituaries; then he'd jet away and run cloak-and-dagger operations in the hot spots of the world. Wasn't that Superman's formula?

"There aren't any boring people; there are just boring questions," Jim Nicholson said once. His background as an investigative reporter was his ace as an obit writer. He'd worked for seven or eight different newspapers, covering dramatic stories—corruption, murder, riots, heroin trafficking, the Black Mafia, and outlaw bikers—and by all accounts, he was a bulldog. Five of his series led three different Philly-area papers to propose him for Pulitzers. By 1982, however, he was marking time at the *Daily News*, running the South Jersey bureau. "I don't know what misfortune befell him to lift him out of that and set him in semi-exile in South Jersey," said Tom Livingston, managing editor at the *Daily News* at the time. The imminent collapse of the *Philadelphia Evening Bulletin* had sent the two Knight-Ridder papers, the *Inquirer* and the *Daily News*, scrambling to capture the *Bulletin*'s readers, and Livingston was charged with snazzing up the *Daily News*. "I wanted only one thing, which was to add a regular obit writer. I knew the readers wanted obits."

In the early 1980s, the obit page was a holding pen for broken-down journalists, and young reporters who needed to be broken in. Nicholson fit the broken-down-journalist slot, Livingston admitted. "I didn't think, 'There's a magnificent writer waiting for the right outlet.' I didn't think of Jim as a terrific writer at all—a nitty gritty reporter, a digger. I just wanted to get him into the paper and thought he could do it well." He took Nicholson out to a fancy lunch, half-expecting him to scorn the offer, but the reporter was sick of writing stories like "A Facelift Begins in Camden."

Livingston said, "To my surprise, he said, 'Sure, I think it's important, too,' and he said it matter-of-factly and without hesitation. I remember him touring the biggest mortuary and talking to the undertakers and finding out what was customary, scoping out the beat, and right from the beginning the obits were very good. He took that weird style the *Daily News* had, which sometimes involved using the second person, and made it his own. And he added something special—he would listen very hard to family and friends and find something in the person's life that was singular. I can remember being at a funeral for the wife of an exec at the newspaper and the son getting up and saying I don't know who that guy is who wrote about my mom but he really nailed it. We heard a lot of that. He would capture something just right about the person. The fact that he had such a wonderful touch with the obits was a surprise," Livingston said. "I just wanted a regular obituary writer. I got the best obit writer in the world."

These new obits created a stir in Philadelphia. Four years after he launched the page, Nicholson received national attention when ASNE, the American Society of Newspaper Editors, gave him a Distinguished Writing Award for his "richly detailed, colorful obituaries of ordinary Philadelphians," the first time anyone can remember ASNE honoring an obituary writer for his writing. The *Daily News* bragged that he had "raised the usually routine job of obituary writing to the level of folk art." Nicholson began to receive invitations to speak to newspaper editors and writers interested in publishing their own ordinary-citizen obits, so he put together do-it-yourself "obit kits," bundles of advice and lists of characteristics to ask about and samples of his stories, and sent them across the country, Johnny Appleseed planting the seeds of the democratic obit. By the time he retired, in 2000, dozens of newspapers were publishing feature-style obituaries of their own "Moose Necks" and "Boss Hogs." Robin Hinch, now at the *Orange County Register*, received one of the obit kits and went on to influence a generation of obit writers herself.

It's a form that can easily turn sticky, and even the best practitioners vary in quality from obit to obit; they're writing on deadline, cranking out the pieces, dealing with the vagaries of real life and the simple fact that there are going to be some sows' ears in the mix. At its worst, this egalitarian obit is sentimental, clichéd, maybe even condescending. The memorable ones, though, are found gold in the pages of ordinary newspapers, like Jim Sheeler's in the *Rocky Mountain News*, or Amy Martinez Starke's in the *Oregonian*. Starke wrote once about a man obsessed with steam trains: "Like a wildlife photographer chasing rare animals, Sandy followed these rare trains with his fellow rail fans from all over the world, especially in the dead of winter, so he could get great photos when the cold air condensed the steam into photogenic 200-foot-high plumes." Who influenced her? Robin Hinch, she said. Did she know Jim Nicholson's work? "The father of us all," she said.

These days, even legends hide in plain sight on the Internet. "Of course I shall help you," Jim Nicholson emailed back. "Us old retired guys look for any excuse to relive our glories." He worked in a joke about his "startling resemblance to Clark Gable," and he called me "pard," as in, partner. He could see me almost anytime, but it would have to be at the house where he was just putting the finishing touches on a threat assessment for the Athens Olympics for a friend on a Washington counterterrorism task force, and where, as he economically put it, he cared for "my wife (Alzheimer's)."

I found him in Cherry Hill, New Jersey, in a brown-and-yellow split level on a cul-de-sac, or, as some would say, dead end. He met me at the door in stocking feet and jeans and a generic feed cap, label removed, which he kept plastered to his balding head the entire day. He dressed like a cowboy: "I take the hat off last and put it on first. I don't feel dressed without this hat." Trim, sixty-two, maybe five nine, he did have the shadow of Clark Gable in his gray mustache and flashing eyes; maybe it was the smirk.

Nicholson's command post, his headquarters, was in the living room, which had all the conventional components, the flowered couch, the mantel crowded with photos, the TV droning MSNBC, as well as the paraphernalia of home nursing, a queen-sized bed, a wheelchair, stacks of disposable diapers—and a woman in a house dress, clutching a rag doll and singing babble. This was Betty, and he insisted I call him Jim. Jim settled me on the flowered couch, next to Betty and across from the TV and the fireplace. He placed a kitchen chair inches from the fireplace for himself, where he could keep an eye on both of us. From time to time, he lit up a Camel or flicked a remote control that ignited the gas fire behind him. He drank black coffee, and periodically he'd zap my cup of tea in the microwave. Everything was leisurely and deliberate, and he made a point of making sure that Betty and I each had plumped pillows behind us and a good lunch delivered by a local pizza shop. He was an excellent caregiver, and he seemed to relish the work. He was also tickled to have company. He'd spoken about obits often while he was writing, but all that obit talk had been bottled up since his retirement and came tumbling out as soon as I hit the record button on the tape recorder.

"The greatest investigative case I ever worked on will be outlived a hundred years by the worst obit I ever wrote. The obits were easily the most worthwhile thing I ever did in the newspaper business. Maybe anywhere. It turned me from a very cynical investigator into more of a Pollyannish, optimist kind of person.

"Most people, in the early years especially, couldn't imagine twenty inches on their plumber in a major metropolitan newspaper. He wasn't on City Council. He wasn't a gang leader. But he got twenty inches for being a good father and a good plumber. I was in a position to do that. You know how good that felt for me? We do a lot of bloodsucking in the newspaper business. A lot of it. This was one part of the business where I felt I could throw wood back on the pile.

"I didn't invent this. If you go back into the archives, to the 1800s, they used to write long, flourishing gothic-novel obituaries, a real art form. I read an early-twentieth-century obit of a woman with a long description of how at the age of five she dined with President Jefferson. All these things would be included. And we sort of lost this. A little life well lived is worth talking about.

"I didn't sit down and say, 'I'm going to invent an immensely popular common-man obit page.' It sort of evolved. I knew I was going to do average people in this. In fact, the general rule there was, if anybody prominent died, I wasn't involved. The city desk would do the councilman, or Mrs. Astor's pet horse. I used to say, 'Who would you miss if you went on vacation, the secretary of state or your garbageman? You'd miss your garbageman.'

"I had a mission to do, and I did it for nineteen years. And it seemed like—aw, I don't want to get mystical, but I really believe I was meant to do this." He lifted the feed cap and swept a hand up from his brow and back over his pate. "There were several years I tried to get a job elsewhere, and even though I had a lot of contacts, I couldn't get a job anywhere else. It was like I was held in place, with one foot nailed to the floor. I had had a serious philosophical disagreement with the editor of the *Daily News*, so I left the investigative beat and bounced around for a little while. They had me working midnights, police beat, bureaus that I started out of my car.

"Tom Livingston conceived of the page, and I welcomed the obits, 'cause after spending two or three years in the Siberias of the *Daily News*, it was a chance to get out from under the city desk, to call my own shots. That's all I ever ask in life. Just leave me alone, tell me what you want done, and go away, and I'll do the rest. A lot of its success comes from the editors not paying attention."

He fetched Betty the rag doll that looked like a rugrat and fussed over her for a minute. "How's that baby, that baby doing good?" She had been chattering to MSNBC for an hour, oblivious to Jim's disquisition on the annoyance of city desks, as well as the news crawling across the screen that Alistair Cooke had died. "I was separated for eight years from this woman, not a good marriage, doesn't matter whose fault it was, and then she got ill about six years ago, and I came back to take care of her," Jim said. "I wasn't doing anything else," he said with a shrug. "I'm not as good as the people I've written about in the obits." As his friend, Lonnie Hudkins, a fellow journalist, explained it, "Jim doesn't leave his wounded behind."

Marriage was the nail in his foot. He was forty, he could type fast with two fingers, and with a growing family (she had been a widow with two boys, then they had a boy of their own), he couldn't afford to walk away from the newspaper business, its paycheck and benefits and pension.

The column, as Jim and his coworkers called his domain on the obits page, captivated the city. The *Daily News*, the people's paper, with Jim's streetwise obits, found its way into every corner of the Philadelphia market, blue, pink, and white collar. The page "was required reading with a lot of people in Philly that would surprise you. The U.S. attorney wrote me a fan note; he had to read the obit every day. There was a federal agent who was a fan. If he was at breakfast in a restaurant, he'd have to put the paper down because he started to cry."

Part of the obits' appeal was in the subjects. "Early on, when I was first starting, I'd just go to the death notices and pick one out. The tipoff would be—'all members of the snake charmers society are invited to the funeral.' Well, obviously I have to know who this is. Another time, I saw the name Slim Lomax, and I thought, that is a great name, because in the movie *War Wagon*, Kirk Douglas's character is named Slim Lomax. And I called up the Lomax family, and it turned out he was the guy who couldn't stand to see a car hood up, he had to get under it. If he didn't even know the person, he'd have to go get under the car hood and find out what was wrong.

"I was able to identify with almost every person I talked with. At some point in my life I had either done that work, been in that situation, or knew people who were. You have too many newsrooms where the writers don't look like, dress like, eat like, or talk like the people they're reporting on. How much in-depth reporting can you do if you don't know what you're looking at? If I was talking to a construction worker, I'm a cement finisher. When I was fourteen, I was doing concrete work with my uncle. So I can finish cement and put in a driveway. I worked on the docks when I was in college, loading trucks. I unloaded flat cars of rocks and threw gravel on roads behind a tar truck. I worked in the oil fields in Mississippi in 1963. I was a car salesman. I rodeoed a bit, so I knew those type of people. I played football in college. I worked as a private investigator. I worked for police intelligence."

That was quite a résumé, and it didn't even count the seven or eight newspapers. Why so many jobs? "Off the record? I was a drunk, a bad drunk. I haven't had a drink since 1968." I measured his direct gaze, his black coffee, his cigarettes, and the gift for straight talk that served him so well as an interviewer; sobriety was his style. (Later, he gave me a "007 license" to include the alcoholism in his story. "It might help some confused juicer out there.")

A workingman who had done lots of different kinds of work, a flawed man with an empathy for humans in all their complexity—and a white man who happened to write comfortably about African-Americans. His obits of church ladies and partying janitors were sympathetic, even intimate.

"I lived for years with my grandmother on a farm in Mississippi. Your basic lower- and middle-class southern white eats and does a lot of things the same as the black. And a lot of your blacks in the northern cities came from the South. They were still from Alabama even though they lived in North Philadelphia. So I would talk to an old black guy whose dad died, about how his dad would use a switch on him. I'd say did they ever notch the switch? Because a parent would make the kid go out and cut their own switch, and some of them would notch it so it would break when they were getting switched. Little things like that, a water dipper on a shelf of the porch, or brush brooms—these were some rural southern things that helped me identify very quickly and ingratiate myself with the black families."

Jim's father had been from Mississippi; his mother was from South Philly, though not Italian, and now lives a few miles from the house we were sitting in. Jim looks after her, too. "I was a service kid. My father was a Marine master sergeant. I went to ten schools by the ninth grade. I was generally left to my own devices and kept my own counsel, for better or worse. That's just who I am." He spent five years in the Marine Corps Reserve. "When our marriage went south and the newspaper job was in the toilet, I decided to go back in the service, and I was given a direct commission as a captain in the army. I put all my efforts and spare time into the military and started going away for months at a stretch—Central America, setting up intelligence units in Panama, force protection in Honduras, the Mexican border, California, Washington. I would go away every year for three or four months and sort of recharge my batteries."

The *Daily News* let him work an eight- or nine-month year on the obits desk, and the rest of the time he ran away from his wife and threw himself into counterintelligence adventures, running the surveillance on Noriega and preparing for the invasion of Panama, fighting the drug war, and who knows what else, though he did allow, "I was on the wrong side several times." He showed a fellow reporter, Leon Tay-lor, how he did the obits interviews and advised him to "be himself" when writing them up, and when he retired from the *Daily News* in 2000, he passed him the reins. Now he and Betty live on his two pensions. "It all worked out," he said.

Betty's name, by the way, is actually Betty Jo. Called Jo during their married life, she became Betty when Jim returned, because he says she's a different person now, "the daughter I never had." He tells me in passing lots of details of her care, almost as if he were narrating his life. For instance, he described how he inflates a wading pool twice a week to bathe her, then drains it with a hose through the garage. Then he said, pointedly, "I'm very happy with where I am, and what I'm doing, and this mission right here. It's a mission, not an in-between, or a stopover, or a temporary—I'm prepared to do this till I drop dead." Later in the afternoon, when her singing got particularly noisy, and he had to cup his hand around his ear to hear me, sitting a few feet away, he started feeding her peppermints to buy five minutes of quiet. "That's a trick you can use when you get in this situation," he said. After he lifted her from the chair to transfer her to a wheelchair for a visit to the bathroom, a lift that required a professional hold and some strength, and they came back ten minutes later and he reversed the operation, he said poignantly, "We never knew each other at all."

The obit writer knew more about the strangers he memorialized than he did about the woman he lived with. That hour on the phone with the survivors was intense. "You are suddenly given an intimate entree into people's lives," said Jim's friend Ralph Cipriano, whose obit-writing tenure at the

Philadelphia Inquirer paralleled Jim's for several years. "You get to hear all kinds of intimate things that people would never normally tell you."

It's a matter of timing. Jim said, "I used to think of this correlation—right after you break an arm, nature, God, will numb you. You can reset your own bone. You won't feel the pain for fifteen minutes, twenty minutes. Right after a death, you can do certain things that you won't be able to do twenty-four, forty-eight hours later. So there's a window. And I could talk to people before it sunk in that Dad wasn't there, or their sweetheart, or their uncle. And if I could catch that window, the emotion wouldn't dilute or distort or close. I'd ask about womanizing and everything. There was very little offbounds. I wouldn't end an interview till I felt I knew the person. And it would click all at once. I would know intuitively that I had it. It would be so exciting hearing them talk, my gosh." He has saved a box of letters of gratitude from the families. Invariably, the survivors and sources write, "I felt so much better after I talked to you and got it out. . . ."

"I had to give this job everything I had. You don't get jaded or careless or lazy if you're an obit writer, because it may be your five thousandth obit, but it's his one and only obit. You can't rest on your laurels because the next family, they don't care what you wrote before. They just want to know, are you going to give them a full-blown Nicholson or not? The good obit writer actually becomes part of the grieving process. A good obit writer takes his place in the chain with the funeral director, the minister, rabbi or priest, the family member." Jim was adamant about that: he wrote for the family. "I was their advocate at the paper, I fought for their space and their dignity, got their picture in. I got paid by the paper, but my loyalty was to the family." He used his vast experience as an investigator not to verify information like their military service—he took the survivors at their word—but to dig out the details that explained who they were.

"It got easier as the years went on because they knew I was looking for quotes, idiosyncrasies, characteristics, so I hardly had to ask any questions. They would have an old obit in front of them, and they would go right down the line. Or I'd ask what would make him mad? What would set him off? He didn't like people walking in front of him when the TV was on, but nothing else bothered him, the house could catch fire. These personal characteristics—we're all different as our fingerprints. And so I'd look for those fingerprints that would make him unlike anybody else in the world. What we call them in intelligence is signatures. When you do counterterrorism work you look for signatures after an incident or an explosion."

What are your signatures? I asked, just to hear him say, Never seen without this dark blue cap, or I take my time smoking my Camels, or I sound like I have a mouthful of false teeth, which I do, or I say I'm a loner, but I can talk to just about anybody. "I don't know. I have no idea," he said, looking dumbfounded. "I don't do much self-analysis. I'm not a complicated guy. There are not many moving parts here. I'm a pretty simple man."

"Who would you miss more, the secretary of state or your garbageman?" and "You have too many newsrooms where the

writers don't look like, dress like, eat like, or talk like the people they're reporting on" are part of Jim Nicholson's basic
obits spiel. He's delivered the spiel to fellow journalists and
students of journalism and even to Terry Gross on NPR, and
he appears in any number of journalism textbooks—and because he's done this before, both as interviewer and subject,
he couldn't help commenting from time to time on the talk
between us in his living room. "I don't have that in any other
interviews," he told me several times, or, alternately, "as I tell
people when I have people to tell . . ." After he said, "It may
be your five thousandth obit, but it's his one and only obit,"
he threw his head back, ball cap and all, and mocked himself:
"Nicholson, you are so fucking quotable!" All that was part of
the business, setting out pieces of pipe and elbow joints and
pointing out the difference between copper and nickel. He
saw himself as the equivalent of the plumber he wrote fifteen
or twenty inches on, a man with his tools who went in there
and did the job. The beautiful synergy that occurred when he
memorialized ordinary Philadelphians at the end of the twentieth century was a mystery. He knew it when he heard it; he
knew it when he saw it—but "It's almost like I knocked over a
couple beakers in the lab by accident, and the chemicals
mixed, and there was this glowing stuff looking at me," he
said. He had gone into journalism because he had a knack for
it. It was easy for him. He has no interest in writing his own
story, or in writing anything anymore. I was welcome to take
what I wanted.

"I don't know why I went into the newspaper business. I
basically don't like to poke into other people's business. And
then I became a reporter. Writing came easy to me, and that's

why I gravitated to it. It was a gift. I don't miss the business. Don't miss it at all. I didn't enjoy writing the last twenty years. I wrote just to feed my family."

He thinks he was made to be an intelligence officer, and he regrets getting into that business too late to have a real career in counterterrorism. An astute guidance counselor could have changed the course of obit history.

Except for the Bible, he says, he doesn't read books. Whose writing style influenced him—Michael Herr, who brought the grunts serving in Vietnam to life? Norman Mailer? Pete Dexter, the National Book Award–winning novelist who wrote a column for the *Daily News* during Jim's era? "No, I didn't read those guys. I read Pete if somebody said he had a good column that day." Ecclesiastes was Jim's influence. "Only six and a half pages, but all the wisdom of King Solomon arranged in his last will and testament." Jim doesn't read newspapers either, and hasn't for years. He gets his news from television and the internet. "It's twenty hours old by the time you read it in a paper!" he said. Newspapers are over. "They had a good run for three hundred years, but now they're dying." Part of his job as an obit writer, he felt, was to bury the newspaper readers.

The only writing he does now is pro bono intelligence briefs for friends in the business, and every once in a while an obituary for an old friend, like the syndicated columnist Tom Fox. "Tom and I worked together as nightside reporters in 1969 on the *Daily News*, and Tom was one of the great writers of his day. Tom's last seven, eight, nine years, whatever, his health deteriorated. Well, his writing did, too, and he was subject to some ridicule. I decided when Tom got in a bad way and was dying, on the chance that I would be off with the army and he'd die off-cycle, when I wasn't at the paper, I didn't want him to fall into the hands of the 'new jays,' the New Journalists. He wouldn't have gotten the tribute he deserved if you only looked at his last ten years. So I wrote Tom's obit ahead of time, to keep him from falling into inexperienced, shallow hands.

"His wife said she liked it better than the one the *New York Times* did, which didn't surprise me. I could beat them on a good day. Fox was from New Orleans, and he really learned his life and his trade on the levees from the blacks—fascinating man. *'As a boy on his bicycle, he followed the jazz bands in the black funeral processions snaking their way to the cemetery. . . . He grew up with the riverfront minstrels, smooth-talking con men, sidewalk buck-dancers, shady ladies and street-corner evangelists.'* Fox became brittle and embittered in his last years, and that's not the picture I wanted the world to be left with, because Tom Fox could blow them all down. *'Fox chronicled the successes, sins and simple meanderings of people who went by such names as Midgie the Barber, Sonny the Bum, Hoagie Joe. . . .'"* Jim quoted Fox on writing four and five columns a week: *like trying to satisfy a nymphomaniac,* and softened his relationship with the new generation with the reminiscence of a coworker: *"For a while we called him 'Fagin,' after the Dickens character who taught all these young kids how to pick pockets for him. He was teaching us the rudiments of news writing: . . . Ironically in his last years, Fox would become somewhat estranged from many of the younger reporters. . . . He said they were too wrapped up in their careers and themselves to record what was really happening in the community."*

By then, Jim was leaning forward, holding his Camels between his thumb and first two fingers and jabbing with it. Maybe it was a style he picked up in AA, or the army, or as a police investigator, I'm gonna give you the lowdown, the confidential; Philip Marlowe. Even though we were sitting in the suburbs of New Jersey around a gas fireplace, talking about the seedy streets of Philadelphia, and our troubadour was a quavery woman in the late stages of Alzheimer's, I had the feeling we were sitting around a campfire out on the open range. He was Ordinary Joe, an old cowboy, telling the greenhorn how it was: There are people vanishing, good people dying out there, and the obit writer's job is to shepherd them out with dignity.

"You know who influenced me? Burr van Atta." Burr was the obit writer for the rival *Inquirer* when Jim was still on the city desk. He had displeased his editors and been punished with the obits job, and he used it to bring the style of the polished *New York Times* and *Washington Post* obits to Philadelphia. "He was a man who never got the credit that was his due, but he invented the modern obituary page in Philadelpia. The details! If somebody smoked, he'd tell you what brand, and how much; if he drank, what kind of beer it was. And he'd put it all in context, what the Battle of the Bulge meant, race car driving, Mayan archeology—he knew so much.

"We always got together two or three times a day for coffee. We'd meet down at the coffee machine, and sit in the little lobby down there. We were a fixture there, sitting in the corner near the elevators, for maybe fifteen years, a few times a day. We spent maybe twenty minutes talking about everything, and we made sure everyone got taken care of. The *In-*

quirer had a forty-eight-hour rule, which was ridiculous, but if you waited longer than forty-eight hours they wouldn't run the obit. Well, I'd run it if it was a week old, it didn't matter. He had space, and sometimes we would trade obits to make sure families got taken care of. Yes, Knight-Ridder owned both papers, but they were fiercely competitive, and still are. And there was one editor who didn't like the idea that we would trade obits, but we did it to take care of the families."

The sensational success of Jim's obits ended up costing Burr his job. "They moved him to put in some promising young writers, some flashy writers to compete with me. They had to hire three people, put three people on his beat when he moved. They started doing a lot of common-person obits. The *Inquirer* wasn't used to getting knocked up." Sports were one thing; the *Inquirer* editors expected to be beat there, "but I bothered them."

As I was leaving, Jim gave me a stack of his best obits, 135 of them, the collection the *Daily News* almost published in book form but didn't. Near the top of the stack was Burr's. "I went to bring him the Sunday paper in the hospital in Camden, where he'd gone for routine back surgery, and the nurse said he'd died of a blood clot an hour earlier. He was my hero in obit writing."

Jim also gave me old memos, citations, letters from the families, contact numbers for his old coworkers. He had refused to let me pay for our lunch, or bring dinner. "Should a guest in my home ever subsidize his or her own visit to me, a shadow would be cast upon my tribe lasting unto six generations." What he wanted was for me to remember his old friends, like Burr van Atta, and Lonnie Hudkins, an eighty-year-old writer for the *Buffalo Evening News* who could make the obit of a woman who spent her life looking out the window fascinating, and Ralph Cipriano, "who often bested me," and Leon Taylor, "who did an outstanding job" taking over for him. Jim wanted to lash them all to his sled.

The only gift he accepted was a DVD of *Lonesome Dove* I mailed him. I hadn't seen the series for years, but somehow it reminded me of him—the cowboy in the counterterrorist. I'd forgotten it was also an obituary story. The cowboy promised his dying friend that he'd bring his body home, and so he did, over three thousand miles of frontier.

7

The Egalitarians

If you want to see where Or-
dinary Joe has led us, pick up any of dozens of papers in the
United States and get to know some of the regular people
who've just departed. A number of the big-city papers with
serious obit pages, the *Atlanta Journal-Constitution*, the *Washington Post*, and the *Plain Dealer*, run several different kinds
of obits. In addition to the daily news obituaries, the editors
give weekly column space to something called "A Local
Life," or "Life Story," where a narrative driven primarily by
sentiment evokes the world of the ordinary citizen. The obit
writers at the *Washington Post* take turns writing A Local
Life; some of these are marvelous and original; some, not. A
few have been unforgettable: one of a woman who contracted Alzheimer's at a fairly young age but had a husband
who was determined to keep her life normal; one of a woman
who fell in love with her female tae kwan do instructor, and
the instructor's husband, too, and moved in with them. And
if there's any question that these indicate the direction of the

American obit, you can look to the American Society of
Newspaper Writers. ASNE recognized Jim Nicholson in
1987 and in 1988 gave a nod to Tom Shales of the *Washington Post* for his obituary tributes. Then they dropped the
obituary-writing category altogether—until this past year,
when they chose Alana Baranick, a writer of life stories for
the *Plain Dealer*, over two news-obit writers, Adam Bernstein, the Kid from the *Washington Post*, and the *New York
Times*'s Margalit Fox. It was a case of choosing the apple over
two oranges, because both Bernstein and Fox are outstanding writers. The judges cited Baranick, whose winning obits
included send-offs for a woman who raised bunnies, a seller
of orthopedic shoes, and a priest's feisty old housekeeper, for
her "richly textured portraits of everyday folks who become
extraordinary through her words," in a citation that echoed
Nicholson's.

*Clementine Werfel blessed priests at St. Joseph Catholic
Church in Strongsville with heavenly desserts, memorable
meals and seemingly miraculous coffee.*

*The retired parish housekeeper, who died Aug. 2 at 96, rou-
tinely walked around the dining table in the rectory, offering
coffee to each priest.*

*"Would Father like regular or decaf?" the 4-foot-something
Werfel asked them one by one.*

*Regardless of the priests' individual preferences, she filled all
their cups with coffee from the same pot. The coffee drinkers
silently accepted what they got, as though Werfel really could
turn regular coffee into decaffeinated, much the way that the
biblical Jesus turned water into wine.*

I never know what to call these obits. Ordinary Joes, perhaps, or common-man obits, though both tags exclude women, and half of their point is to get at the lives of the hidden women in this country. I prefer "egalitarian obits," a phrase I first heard from historian Nigel Starck; it captures both the essential Americanness of this kind of obit and its effort to bring women onto the obits page. Three big academic studies have analyzed the modern obits page, Starck says, and all three found that women account for only eighteen to twenty percent of the obits, even in contemporary New York. Minorities continue to be grossly underrepresented. Gay Talese had it right forty years ago: on the obits page, "women and Negroes hardly ever seemed to die."

What does it take to be a good obituary writer? Reporting skills, life experience, and something I can't quite pin down—an ability to weigh someone's life and accomplishments historically, in the context of the times. A good obit writer has to communicate the significance of a person, a place, an era. Sense of humor? It's a good survival skill, and it adds a lot to the pleasure we take in some writers' work, but I don't think it's essential. Neither is empathy. As for "style," that's also optional. What it takes to be a good obituary writer is an ability to write well, to capture a person with economy and grace, and work in the hurricane of emotion that swirls around the newly dead.

Agate, population 70, is one of those towns that people describe as "blink and you'll miss it."

Lois A. Engel loved living in the blink.

(by *Jim Sheeler*, Denver Post)

She *loved living in the blink*. I needed to meet the person who wrote that.

Jim Sheeler might have been one of those careerist "new jays" scorned by the old guard, thirty-six years old and smooth shaven, serious about his writing, a nice rep tie complementing his white shirt and khakis, and such nice, straight, white teeth, a whole upwardly mobile past could be read in their gleam. But don't assume anything about him until you hear the story of his grandmother, Betty Sheeler, who moved to New Orleans when she was seventy-two, opened a private post office on Royal Street in the middle of the French Quarter, and welcomed all customers, HIV negative or positive. Sheeler remembers sitting around with "Miss Betty," as the locals called her, eating boiled blue crab out of newspapers and visiting with the fascinating people she had figured out how to meet: the jazz musicians, or the woman called the Duck Lady, who skated around with a duck in her arms, cursing. No matter what Sheeler did with his life, he wanted it filled with such original characters.

Sheeler grew up in Houston, then studied journalism at Colorado State University. There he worked with Garrett Ray, who had edited the legendary Littleton, Colorado, *Independent*, and believed passionately in the importance of "regular folks." (His predecessor at the *Independent*, Houstoun Waring, had written 13,000 obits of ordinary Colorado citizens before he died, at ninety-five.)

After graduation, Sheeler decided to stick around and work for papers like the *Daily Camera* and the *Boulder Planet*. One thing he couldn't handle was elbowing his way past other reporters, competing for stories. He figured he could avoid that scene if he wrote about people who had never been written about before, people nobody else wanted to talk to. Eventually, he found steady freelance work for the *Denver Post*, and early in 1999 the *Post* announced the debut of a new column.

People die every day and their lives go largely unnoticed. Starting today and continuing every Sunday, The Denver Post will feature one of those Colorado residents. Through the writing of Jim Sheeler, The Post will bring to you obituaries about the everyday lives of extraordinary people across the state.

Sheeler took the bias and goals of the egalitarian obit, kept a lid on the overblown language and the clichés, and used the techniques he had learned in journalism school to report the hell out of people's stories. His writing was cinematic.

The way his relatives figure it, Fred Hughes' life would have made an amazing movie—a blockbuster adventure tale, they say. They already know the perfect opening scene:

A single-engine plane struggles its way across the Gulf of Mexico sometime during the mid-1950s. The leading man tries to help navigate from the co-pilot's seat while his wife furiously works at a makeshift pump in the backseat. She is their only hope to keep from running out of gas.

The cinematic influence wasn't just the natural expression of a kid who grew up on television, but the result of tools that came from his courses in broadcast journalism, his major. The overriding imperative was to *go out and get the story*. This

works for him. "I'm one of the crappiest telephone interview-
ers in the world," he tells me. "I go out as if I have a camera
on my shoulder. I almost *produce* my stories. I wrote the obit
of a letterpress printer, and her apprentice wanted to know if I
would like to see her office. 'Yes,' I said, 'and I want you to be
working in it while I interview you.' So he printed her funeral
brochure while I talked to him."

A list of his obits that ran under the title A Colorado Life
sound like pure cornball—FROM A SURVIVOR TO A SUCCEEDER:
AMPUTEE LOST LIMBS, GAINED GRIT; SMALL FRAME HOUSED
LARGE SPIRIT: "LEPRECHAUN" SHEEHY LOVED TALES, TYKES —but
time after time they redeem themselves with carefully observed particulars and unpadded writing.

At some point, Sheeler read Jim Nicholson's obits and said
to himself, Okay, see, this isn't a new thing. Other people
have done this, too.

The *Denver Post* tried to persuade Sheeler to come on staff,
but freelancing was essential to his feeling of independence.
They took the column away from him. The wonderful Claire
Martin replaced him at the *Post*, carrying on the task of writing about the overlooked, and extending it with compelling
obits of homeless men and months-old babies. Sheeler moved
to the *Rocky Mountain News*. There he began writing about
the young people of Colorado who watched the Twin Towers
fall on television and were determined to do something about
it. His obituaries and feature stories about the slain soldiers of
the Iraqi War are valuable as history, sociology, literature, and
journalism. Are they obits? If you take into consideration that
he wants to break as many rules as possible, then, yes, they
are. I recommend reading them in private.

The silver van was spotted quickly as it slipped into town and turned up the old road that everyone here knows, raising questions and concern in its dust.

One of the checkout clerks at the 90-year-old Hayden Mercantile was one of the first to see it, and wonder. In a town of 1,634 people, it wouldn't be long before nearly everyone knew the weight of the news it carried.

The van drove past downtown, past the houses with blue stars beckoning from the windows, signaling that the family has a son or daughter overseas.

It drove past the old feed store and the silos and the history museum in the train depot. It drove past homes with a single American flag and turned toward the place a bit farther out in the country, where the flags sprout like wildflowers.

It was the beginning of the Labor Day weekend, and Sherri Lawton was on the phone when she saw the van driving past all the yellow ribbons and flags, past all the names she refused to forget.

"Hold on," she told a relative on the other end of the line. "There's a silver van coming up the driveway."

She stopped for a moment then added, "I don't know anyone with a silver van."

"Ohhh," she said, "it has government plates."

When the man in the uniform got out, he was already crying.

"There are times I cry during these things," Sheeler admitted. "I bawl all the way home."

He takes his time getting the stories. To his chagrin, the families of the dead soldiers are in demand for interviews, but he refuses to stake them out. He writes a letter to the family expressing his sorrow, not a form letter, and hand-delivers it. About half of the families say they aren't ready to talk. The rest invite him in. "The families want these stories to be told," he says. He hangs out long past the conclusion of the reporting, stays in touch with the survivors, writing follow-up stories for the *Rocky Mountain News*. "I'm friends with all the gravediggers," he says.

Amy Martinez Starke of the *Oregonian* is a storyteller disguised as an obit writer.

Frank Zielony might have lived his entire life as a Polish farmer and brick maker like his father, in the plains of what is now Ukraine. But in 1939, war came. Soviet troops showed up at 7 o'clock on a dark morning in early 1940 and told the entire village—including 19-year-old Frank, his parents, brother and four sisters—that they had half an hour to prepare to leave the country. They were packed in cattle cars and deported to Siberia—among more than a million non-Jewish Poles forced into slave labor camps.

That's how Frank, about 5 foot 4 and 150 pounds, came to be cutting down trees and making railroad ties in sub-zero weather. Food meant potatoes. Sleep meant a bare floor, and taking turns. During those two years, he lost his mother, father and three sisters to starvation, illness and exposure—his mother was buried while he was out in the forest overnight cutting wood. None of their graves was marked.

(Oregonian)

The story of Zielony's survival and immigration to the Pacific Northwest, and his life helping other Polish Catholics survive, was one of those great obituaries that are made to be reread. Dark and gritty, but suffused with spirit, it was written by someone inspired. Amy Martinez Starke read about Jim Nicholson's obits back in the eighties and "made a mental note that I might want to do that someday." The time came in 2002, when Starke, who has been a journalist for twenty-five years, proposed writing three feature-style obits of ordinary people a week. It was "an easy sell. My managing editor had been aware of other papers and knew it was very popular with the readers." The *Oregonian* had also been one of the newspapers syndicating Portraits of Grief after 9/11. After the first week of Portraits, the editor tried to stop reprinting them, but there was an outcry from readers and staff members alike. Portraits stayed. The next step, longer Portrait-influenced obits of the ordinary people of Oregon, was a natural.

The *Oregonian* has always run the equivalent of paid obits, except they give the space away. These free agate-type obits are written up using forms filled out by the survivors, with information about how old the person was, what she did for a living, and so on. Starke and the other Life Story obit writer, Joan Harvey, browse through stacks of these and choose subjects from the forms. Starke is a busy woman; she and her husband have two teenagers, one applying to colleges and one in the band, and her obits are labor intensive. She has to be efficient, but she lavishes time gathering raw information. "My general rule of thumb is no less than five interviews per story, usually closer to ten. The writing doesn't take much time, but the interviews, they can be at least an hour—and, yes, everything by phone." The distance is crucial. She says she couldn't do this for someone she knows.

"Do I feel a mission? Oh, yes, to increase circulation. That's the bottom line. If people would take the paper for one thing, for the Life Stories, I'd feel like I've fulfilled my mission here. If journalists feel they have any other mission, I think they're misguided. Without newspapers we're not going to have any-place to write."

Her businesslike approach is in contrast to her soulful writing. Her ordinary people are complex and emotional, written about from a distance that's hard to pin down. Where is the author? Sometimes it seems she's chatting about the story over coffee, gossiping: "Sister Anne was a card player and could play pinochle for hours on vacation in Lincoln City with other nuns, after morning prayers. She was known for taking a little whiskey neat. . . ." Other times, the author lays out the string of details and biographical information, but she seems to be spinning tales from the spooky distance of a nineteenth-century novel, or a legend:

> Ursula King was considered unruly as a girl. Her father, a strict minister, chastised her for chasing the magnificent police horses through the streets of Lincolnshire, England, and shipped the 10-year-old off to work on an aunt's remote sheep ranch in Australia. A pony she rode to school made up for a lot, but Ursula neglected her chores, and her uncle shot the pony. While she was attending a boarding school in Sydney, a circus came to town. Drawn by the idea of working with horses, Ursula ran away to join the circus.

I loved the Ursula King obit the first time I read it, and I still love it. It reads to me like a short story, or, even better, a myth that happened to be true. Toward the end of the obit, where the story of a life usually winds down with diminishment and disease, Starke places her subject in the sickbed—but it is a sickbed with a view.

It had always been her dream to get a black stallion, and she finally got one in 1987, at the end of her breeding career: a black colt, a 16-hand true-black Egyptian breed Arabian named Nite Magic. Her bed was by the window, so she could see Nite Magic in his pen, and he stood at the fence and stared at her. But she couldn't ride him.

The picture of the dying woman locking eyes with her Arabian horse haunts me, and that harsh last line has the weight of tragedy.

How does she get that effect? "I don't use attributions," Starke points out. Her obits aren't cluttered with phrases like, "according to her nephew," or "his agent announced today," or "as she told the *Oregonian*." It's a simple variation on newspaper style, which nails down an attribution for almost every fact; removing it gives Starke the omniscient authority of a short story writer. She doesn't care if her subjects are good guys or bad guys, sinners or saints, "as long as it makes a good story."

Catherine Dunphy writes almost exclusively about people who've never had their names in the paper before. She brings the real people of Toronto to life in their obits in the *Toronto Star*, firing off quotes from family and friends and great details to build character.

[Eddie Baltimore] had a mojo bag given him by a voodoo priest. He kept red dirt from the Delta in a beautiful coffee-table container. They were beside him as he lay dying of cancer in a bed set up this spring in the living room of his quirky East York bungalow.

[Bernie] Share, 74, was the man who owned a neighbourhood mecca. Officially it was an antique store called Gallery and Innovations; unofficially the cluttered storefront with its wondrous wares tumbling out onto the sidewalk on the south side of Queen near Niagara St. was a neighbourhood clubhouse for the denizens of this neighbourhood and customers from all over the city.

Everybody knew the guy with the English accent in the three-piece suit and the cowboy hat, and lots of them referred to him as the King of Queen St. . . . Share's presence is everywhere in this store, with its bric-a-brac, books, hat boxes, African carvings, brass beds, Czech glass and Raggedy Ann dolls. One of his cowboy hats is hanging on a peg. His chair is still at the front of the store. And his portrait sits atop one of the stacked dressers, overtop the shelf holding his favourite collectibles: the elf salt and pepper shakers, the hula dancers. In it Bernie Share is smiling—and wearing his black cowboy hat.

I've saved dozens of Dunphy's obits. She manages to make Toronto, a city I've never seen, into a place I feel I know. I treasure the obit about the American expatriate who moved to Toronto after the National Guard opened fire on students during an antiwar protest in 1970 at Kent State University, killing four. Roni Chaleff built a community of friends in her adopted city. "She was the glue for many local groups, forging bonds so strong they were transformative," Dunphy wrote of Chaleff's social justice organizations, baby-sitting coop, and women's collective. "Chaleff never strayed from her ideals. She refused to own a house, dress for success, eat meat." When she was diagnosed with cancer, she asked an old friend to be her chemo-buddy. "They packed picnic lunches for the six-hour sessions, 'took stock, reviewed old boyfriends, and laughed. We had such a good time,'" the friend reported. As one of her sons said matter-of-factly after her death, "My mom isn't here, but everybody else is."

Sheeler, Starke, and Dunphy are just a few of the writers whose obits I collect. They combine the virtues of obituarists and feature writers, fashioning emotionally evocative obits without the mawkishness that so often characterizes this hybrid form. I read them and dozens of others on the Web. I read them for a few cents on the newsstand when I travel. Newspapers are ailing, cost-cutting and folding; as Nicholson pointed out, the great age of newspapers is ending. It seems right that their obits are flourishing.

8

Tributes

When I first became obsessed, I was writing celebrity obituaries for magazines. I'd find celebrities on the brink of death, being carried off in stretchers on the cover of the *Enquirer*. That was my cue, "Bob Hope Near Death!" or "Johnny Cash's Last Words ('Wait for Me, June')." I'd go into a frenzy, hurrying to finish the obituary so it would be ready to publish as soon as he exhaled his last breath, racing against time. I researched the stricken one's filmography, discography, romantic history, turn-ons and turn-offs—everything!—with hyperventilant speed, then wrote these long, detailed, passionately felt tributes . . . at which point the near-dead celebrity, like the old dog who had a can of cat food waved under his nose, would spring back to life.

I tried to be discreet. There were guessing games on the sidewalk and in the playground I frequented in my town: Who was old enough, sick enough, famous enough for these mysterious obits my neighbors wondered, but I wouldn't tell.

My celebrities would be plastered over the tabloids, frail, ghostly old men and women being spirited in and out of the hospital. Anyone could have seen me leaving the library or the video store with stacks of old movies, racking up overdue fines on stacks of old books. I smuggled biographical booty into my darkened house. Some of the movies flickering on my screen were so old and dated, they might have been early talkies. Some of the tapes and CDs I played sounded scratchy, as if they had been recorded under a freeway. I carefully hit the mute button when the telephone rang.

I woke up tense every morning: Did my subject die yet? The answer, every morning, was no. Maybe that's what I loved most, that emotional ratcheting-up, writing something charged with a sense of urgency and tragedy, in a secret bubble of time. Celebrity reporting is usually a feeble exercise in dressing up someone's canned answers, but this was infused with drama. By the time I finished writing, I was a true fan. I knew what the world would lose when it lost each of them.

I might have known someone like Elizabeth Taylor would survive the dire predictions of the tabloids. In one of her old photographs, she sports a fresh tracheotomy scar and grins over her premature obituary ("The best reviews I ever had"). I wrote a tribute to her forty years after that tracheotomy saved her from death by viral pneumonia, and after a host of other physical maladies had failed to fell her: a brain tumor, three hip replacements, multiple broken backs, more pneumonia. It seemed, in her late sixties, that her terrible health and all those years of pills and booze would finally catch up with her. Instead, she recovered and made a TV movie, *These Old Broads*. The first million-dollar actress dusted herself off and went back to work, flashing her exquisite world-class jew-els, then tossing them on a sink cluttered with dried squirts of toothpaste and old lipsticked liquor glasses. If you hear she died, don't believe it. By now, I suspect, she's immortal.

I kept Elizabeth Taylor and Katharine Hepburn and Bob Hope and Johnny Cash in a mournful file in my drawer. There was something eerie about my failure to pick goners. Maybe I had special powers, a reverse jinx? This is proof we are doing God's work, my editor said stoically. I was collect-ing stories about the failing Princess Margaret when he called up to bark, Drop everything, Brando's dying! Marlon Brando had just collapsed on a movie set and been lugged to the hos-pital, seventy-something, massive, and massively unhealthy. I plunged myself into all things Brando, from the wildly erratic films to the 1,118-page biography full of minutiae like Brando's friends' reactions to the antics of his horny pet rac-coon, Russell.

Would I have liked to meet Marlon Brando? No, thanks. I didn't want to be toyed with by a celebrity who was famous for, among other things, toying with reporters. I never had a desire to meet Brando or his raccoon, and after wallowing in his oeuvre, including a memorable day at the Museum of Television and Radio in Manhattan, viewing clips of him feverishly playing the bongos, I was relieved to be hiding be-hind an obituary.

But dead, or near death, and from a distance? I loved him. I loved studying him. I developed real sympathy for the guy. I saw him as a hurt child, imprisoned by the circumstances of his childhood. His mother and father were at war for his soul—the sensitive (drunken) mother, the brutish (drunken) father—and he internalized both of them. He was gorgeous in his younger years, and hugely talented, but contemptuous of the fame and flattery they brought him. *Ya still like me, ya losers?* he seemed to be saying each time he gained another fifty pounds, or wasted himself in a bad film, or mooned the tourists. He was his own worst enemy, so contrary that whatever I found to say about him, the opposite also seemed true. Anyway, I had time. Brando, true to form, had recovered in spite of himself, and gone on to appear as a criminal drag-queen in *The Heist*, wearing lurid makeup, a dress, and, according to indiscreet coworkers, a sophisticated earpiece that delivered his lines to him seconds before he delivered them to the camera. Instead of dying, he was busily adding strange chapters to his life, while I watched the leaves fall and the snow melt and the new baby deer run across the highway.

I slid Brando into the drawer with the others. There they slept for several more years, while the magazines I wrote them for either lost interest or, like *Life*, died.

Then Hepburn, Hope, and Cash all decided to go in one summer's glorious, staccato burst: Hepburn in late June at ninety-six; and Hope in late July at one hundred; Cash, a mere seventy-one, hailed the chariot in mid-September. I felt my adrenaline surge each time. No matter how many times you imagine the moment, you still can't believe it—there's a whoosh of energy that leaves the planet. They're history.

Brando, typically perverse, followed a year later.

Anyway, I was not, after all, writing obituaries, as I learned after a few months of disciplined reading. I was writing tributes, literary set pieces that if I'd been lucky would have appeared entombed in the pages of a magazine, along with many glossy photos of the dead star. I read such stories now, after a couple of years dining on newspapers, and I can't believe how fussy they are. I tried to assemble the facts and write something truthful, as far as I knew, but the 200th time I lifted one up and studied it from all angles, I began to notice some of the life had been written out of it. You know how water is? It gets funky if it sits around. It needs to be part of some flowing stream or churning ocean.

I haven't become a purist, but I've lost my taste for the studied tribute. The *New York Times* in its Sunday magazine runs a roundup of tributes at the end of each year. These tributes are a big production. No workhorse prose on short deadline by one of the regulars for these; not even an advance by someone especially graceful. These are the hors d'oeuvres of literature, written by a literary name, and displayed with one spectacular photo apiece. The magazine is a keeper, designed to sit in a pile of glossy detritus on the floor by your bed for at least a year, until the next one comes along. I admire these memorial issues hugely; I turn down the corners of my favorites and study the writers' styles. *Life*, my old home base, used to publish a similar issue in December each year: a theory dressed up in anecdotes and quotes, glittering slivers of details, and sentences that sound like voice-overs at the end of a classy documentary. There's nothing wrong with them, except a bit too much self-consciousness, but I rarely finish reading the whole issue. All those weighty thoughts, hot air in hushed tone.

And, come to think of it, I don't love those long advance obits of the mighty that have been sitting around the morgue and end up running hundreds of inches in the newspaper, either. I can't put my finger on a reason. The prose just seems dead, stuck in heavy, industrial boots, with tons of information shoehorned in. When the great one finally dies, this obit is sent into the big boxing ring of the newsstand to try to knock out the competition. It's a far cry from the slippery puddles of mercury that spill off the desk of the regular obit writers a few minutes before deadline, the lives of the not-so-famous or the famous who take us by surprise when they die. That pressure, and the compression that results, that comes on the heels of death—particularly the shocking death—charges the prose with electricity. There's a living mind trying to capture an essence even as it's being whisked away. Wait a second! the instant obit writer begs. Tell me what she loved. What made him say, There was my life before this, and my life after. Why she got off the train in Chicago . . .

Joe Holley's '04 obit of a man named Kenneth Edelle Foster in the *Washington Post* could only have been written on deadline.

Mr. Foster was working in his office in the Hoffman Building in Alexandria on Sept. 11 when he got word that a plane had hit the Pentagon, where his wife, Sandra Nadine Hill, had worked for 25 years. He jumped into his truck and raced toward the billowing black cloud he could see in the distance, going the wrong way on Interstate 95.

He ended up spending nearly two days and nights helping rescue efforts while desperately searching for his wife. Because he wasn't supposed to be there, a woman gave him T-shirts from the Salvation Army and the Red Cross to wear so he could blend in with the rescuers.

After surviving a deep depression, including a suicide attempt by Russian roulette, Holley writes, Foster sought therapy, moved to Texas, and set up a scholarship named for his wife. He died at fifty-one of pulmonary fibrosis and congestive heart disease. The obit ends:

> *"He could have got over his physical ailments, I believe," his mother said, "but he just didn't want to live. He died of a broken heart. We all know that."*

I've read last lines like that before and rolled my eyes, but this was earned, this swift, rough aftershock of an obit from 9/11. Nobody would have written an advance obit of a fifty-one-year-old man, a civilian army policy analyst. Saved as an end-of-the-year tribute, it would have read as too neat and too pathetic, a ponderous tragedy. It was meant to be the infusion of one man's sorry end into the cradle of the news obit, with just enough there to tell his story. I don't care if the verbs aren't perfect and not every clause sings. I'd take it any day over a full-throated tribute to Bob Hope.

There is one place to read obituary tributes with that vital spark. Hold on—I'm walking up Ninth Avenue, squeezing past a messy corner of trash barrel, news kiosk, souvenirs spilling from little holes in the wall, on the way back to the *Times*'s building from my elegant lunch with Chuck Strum. It's an afterthought. We can hardly hear each other in the traffic and street noise. Besides the *Times*, who runs your favorite obits? He doesn't say anything for a minute. Maybe he doesn't hear. No, he's considering while we pick our way to clear sidewalk. As soon as he thinks of it, his face relaxes into a smile. The *Economist*! he says.

I didn't even know what it was when I started my research, "more newspaper than magazine," as one critic wrote; an influential weekly, *Time* with a supersonic thrust and a history that went back to 1843; a British publication with half its circulation in the United States. Its articles are anonymously written and smart, worth a dozen consultants, they say. And every week, on one beautiful page, a single character is sent off in memorable style, an obit dressed lightly in essay form: "Canaan Banana, clergyman, politician and rapist," or Janet Frame, the New Zealand novelist who was "a chronicler of mental turmoil," or Filiberto Ojeda Rios, Puerto Rican revolutionary.

The magazine, smart and alert as it is, had noticed the revival of obits in the London newspapers. Robert Cottrell, a journalist from the foreign department of the *Independent*, carried the bug for good obits with him when he became feature editor of the *Economist* in the early nineties. He ran a wonderful piece in late 1994 about the phenomenon of obits as entertainment that, he wrote me, "helped to persuade many of my colleagues that the subject was worth exploring." (The piece ran anonymously but was written by Martin Vander Weyer, who contributed obits of business figures to the *Daily Telegraph*.) By the mid-nineties, Cottrell and his anonymous contributors were in stylish competition with the London dailies.

Cottrell was replaced on the beat by the novelist Keith Colquhoun. "Keith made a name for the *Economist* obituaries;

he'd take people who were almost unknown and had done something like invent a new sort of flower poem, and then he'd talk about flower poets for most of the obit. Beautiful little pieces, not really an obit of that person." So spoke Colquhoun's replacement, the biographer Ann Wroe. About 80 percent of the dazzling obits that have appeared in the *Economist* since late 2003 are written by Wroe; she edits the rest of the print-edition obits (the American team that runs the *Economist*'s website run both their own obits and obits from the magazine).

I got all this on the fly. "I am writing this from a hotel in Bratislava," Cottrell's email read, and Wroe was run ragged trying to track down some of her earlier pieces. "Weirdly enough, we have no run of the printed issue anywhere in the building!" she wrote in exasperation.

Wroe looks like (and is) a sensible Englishwoman of fifty-four. She wears sensible shoes, a dark skirt just past her knees, and a tucked-in blouse, and unfussed-over gray hair frames her long and lovely face. But she's lit by the fire of her passions. She's worked at the *Economist* for almost thirty years (in an office building off Piccadilly—"You can't miss it. It's on the street that ends in a palace"), but honestly, she says, it won't merit but a line in her own obit. "I really haven't been here. I've been elsewhere, thinking, writing my books." She has four eclectic books to her credit, and the windowsill of her cramped, shared office is stacked with research on Percy Bysshe Shelley, her latest subject. She talks about writing obituaries with the ardor of a mystic.

"I like to get into the mind, into the shoes of the person I'm writing about. I like to get as far as it's possible to get into somebody else . . . almost a merging of the character. You're only dealing with them for two or three days, but it's very intense. To see through all the information to the essence, to that moment in a life that's the key or the facet of the personality that is the key to the whole person. To try and see the world."

She chooses a subject by Monday morning, after consulting the *New York Times*, Nexus, and the Blog of Death, a weblog that posts tributes to the recent dead, and by Wednesday, Wroe has handed in her piece. "The picture is terribly important in the obit, I always find. Sometimes I can't start to do it until I know what the picture is." The visual inspiration and the speed with which these obits are written might explain why they have an immediacy and spirit not often found in magazines. She seldom falls back on advance obits. "We have almost none in stock. We have about four. I try to get them out of people in advance, but journalists are bad about that deadline, and some of them say, I won't know what to say till it happens. I've been trying to get Maggie Thatcher out of the political editor, because she is somebody who, if she died on a Thursday morning, we would have to remake the whole paper, for goodness sake—and he won't do it.

"We did an obit of an Indian bandit a while ago, one of my favorites; it was done by our Delhi correspondent. And this guy was the biggest bandit India has ever known—he killed four hundred people and five hundred elephants—and we got quite a lot of letters saying, You must not write about this man. He's no good. You must only write obituaries of people who are worthy. This is a fairly prevailing sentiment, actually.

"We're not very reverent in our obits. They have a tradition of that in England. I think the best obits are the ones in the *Daily Telegraph*. They set the standard. Their obits of minor aristocrats and war heroes are the greatest. There have been some classics where you have a dissolute person who has done nothing good in his life and has just been crazy—and he'll be treated with the most wonderful seriousness. Sometimes the *Telegraph* will do an obit, and I'll just think, there's nothing more to be said about a person. I would like to have done it myself, but no, it can't be done any better than that.

"It's a writerly job. It does need very good writing, this page. That sounds awfully boastful, but it's a form of literature."

Wroe calls the book about Shelley she's writing in the evenings and on weekends "an alternative biography. I'm trying to write about him as a poet, not as a man. And the narrative drive of the book is the progress of his soul out of the body. It's mind-blowing—I just don't know if my mind is good enough." It's one of those things she attempts "just to see if it's possible. I got very keen on biography because I wanted to change it. I wanted to stretch the form. I think of it as a way of capturing souls."

There is in her intensity and language something Catholic (and, in fact, among her other jobs, she writes a regular column for a Catholic newsletter). Also, she's at home with guilt. "I did an obituary on Nigel Nicolson. He was allowed to see his mother for half an hour a day. And she would talk to him but she was still sitting at her desk with her pen in her hand, and she would only turn around slightly to talk to him. And I thought, that's me. I've caught myself doing that. The boys [she has three sons with her husband, an actor] come in and lie on the floor, and I do chat, but I've still got my pen in my hand . . ."

But guilt doesn't stop her from reveling in her work and its riskier challenges. "I do, I have fun!" she enthuses. She relishes the writing assignments that involve a walk on the dark side, particularly with writers like Hunter S. Thompson. The photo that inspired his obit is a shocker: Thompson leers as he points the business end of a pistol at the photographer. Readers look directly into the barrel.

There were always way too many guns around at Hunter S. Thompson's farm in Woody Creek: .44 Magnums, 12-gauge shotguns, black snubnosed Colt Pythons with bevelled cylinders, .22 calibre mounted machine guns. He also kept explosives, to blow the legs off pool tables or to pack in a barrel for target practice. His quiet bourgeois neighbours near Aspen, Colorado, complained that he rocked the foundations of their houses.

Explosions were his speciality. Indeed, writing and shooting were much the same.

Wroe ends the obit with a quote from Thompson about Hemingway's suicide, in which Thompson sympathizes with the old writer's despairing thoughts as he lost control of his world: ". . . finally, and for what he must have thought the best of reasons, he ended it with a shotgun," Thompson had written.

Writerly, without a lot of hollow bullshit—that's the secret. One of the best things about Wroe, a shopkeeper's daughter who studied at Oxford, is her lack of pretense. She's amused that I can't figure out what class people are from in England. Did I notice the man who just poked his head in the office? He had a nice suit, I thought. She laughed and told me anyone else here could look at him and see baronial estates, titles, hunting dogs, the whole palette of "upper" life. "Also, they're taller than we are," she said. Wroe envies Americans' lack of pretension. Her favorite thing is to fly to the U.S., rent a car, and head west, a rubbish country-and-western station on the radio. "You can't imagine how liberating that is." I imagine her speeding over the Appalachians to the tune of "Save a Horse, Ride a Cowboy," stretching the boundaries of her personal biography.

Wroe signs a copy of her book *Pilate: The Biography of an Invented Man* for me before I leave. Almost nothing is known about the man who condemned Jesus to crucifixion, so Wroe wrote the story of his times, weighing his myth as it has been constructed through the ages. Working with only scraps of evidence, she pieced together a cohesive, dramatic story. No one, for instance, knows Pilate's first name (Pontius is a reference to his tribe, not his given name). With the poetic delicacy that distinguishes her obituaries in the *Economist*, Wroe wrote, "his mother bending over his cradle would have whispered a name that history has rubbed away."

9

The Four Horsemen of the Apocalypse

THE OBITUARY CAPITAL

The week before I flew to London, the fascinating Lady Penelope Aitken had died at ninety-four and was remembered affectionately in all four of the London papers that regularly run obituaries. She had most recently stood by her disgraced politician son, Jonathan Aitken, who had had to serve an eighteen-month sentence for perjury. It was "just a blip, an awful blip," the *Guardian* had quoted her. The *Daily Telegraph* noted that her loyalty to her son had been tested before: "When it emerged that he had fathered a child with Soraya Khashoggi, the former wife of Adnan Khashoggi, she welcomed her new grandchild into the family."

Pempe, as she was known, had been a great beauty, a bewitching woman who charmed kings and princes, and had affairs with all sorts of men. The *Independent* dug up a delicious quote of hers, referring to her marriage to a staid journalist and "the deep, deep peace of the double bed after the hurly-burly of the chaise-longue." According to *The Times*, "Even after the age of 60 she continued to attract younger men, some her son's contemporaries," including one who "eventually moved into her life, remaining her companion and walker, and cooking exotic dishes at her parties." There were long queues at her deathbed, where the gentry and hoi polloi and some of the felons her son had befriended in prison came to say goodbye (one of his former fellow inmates had become her chauffeur).

It was hard to say who had the best obit—really, you'd want to read all of them—but the *Telegraph* and the *Independent* vied for the most exotic details.

When [her son Jonathan] took LSD to write up the experiment for the Evening Standard, *[Lady Aitken] was there to record his hallucinations: "He had the most vivid and terrible dreams . . . he thought he was in Australia, he could see the future and there were just seas of blood," she recalled.*

(Daily Telegraph)

She was not in the least snobbish. Mr Ray from the nearby gypsy encampment was a regular lunch guest . . . partly because Pempe rather sympathised with his view of the proper way to die: "I wanna be lying in a ditch with the beer running outta me mouth."

(*by Christopher Bland,* Independent)

Do they make people like this in the United States? If so, I can pretty much guarantee we don't read about them in the obituaries. Pamela Harriman didn't come close (though she and Lady Aitken probably hurly-burlied through some of the same boudoirs). I have the feeling that London, gossipy and contentious and cutthroat as it is, unites to embrace its characters. There goes another one, each obit seems to say, whose likes will not be seen again. All witnesses to the accomplishments and antics of the deceased must be gathered and debriefed. Only the most thorough and entertaining and true account of her life can be thrown down, like a royal flush, on the table. A four-newspaper salute to the ones who succeed in shrinking the world to a neighborhood.

All of which is to say that when my husband and I met the freelance obit writer Tim Bullamore outside the Houses of Parliament and strolled past Parliament Square and Westminster Abbey (Bullamore's five-year-old daughter skipping merrily ahead), we passed a hearse parked by the Abbey entrance. Bullamore is trained to take in the details of death. He tilted his head to read one of the labels on the flowers. "Why—it's Lady Aitken!" he said with familiar pleasure.

We read the obits. We all knew Pempe Aitken, bless her vivid soul.

I couldn't wait to get my hands dirty in London. Later for those thousand-year-old churches and palace guards; the idea that you could go to a newsstand on any corner and buy a dozen newspapers is what makes London special to me. I had never been to the motherland. Friends gave me guidebooks, and lists of all the monuments and museums and markets and restaurants I had to visit. They said I had to see Buckingham Palace and ride that giant Ferris wheel thing on the Thames, and I was a fool if I didn't hang out in the National Portrait Gallery and the Tate. But instead I spent the first morning of my pilgrimage at the newsstand in a haze of pleasure. Then I drifted back to my room with the four papers at the center of the obituary world, *The Times*, the *Guardian*, the *Daily Telegraph* and the *Independent*—the Four Horsemen of the Apocalypse. I had fondled these papers before in New York libraries; occasionally I spotted them at the odd newsstand for three or four or five dollars each and emptied my wallet. I had never owned them all on the same day, or spread out the four obit sections side by side, or devoured them while the sounds of the London streets blared through the open window. They covered all the surfaces in my tiny room—and they were beautiful. A palace is just a building. The British obit pages are works of art.

No city in the United States can offer this many newspapers, snarling and fighting over their version of the news of the dead, throwing down obituaries like they were dice in some ultimate craps game. In 1986, the *Independent* was about to stride onto the London scene with illustrated obits designed to challenge the centuries-old *Times*. *The Times* beat them to the punch, and within weeks of the appearance of the *Independent*, buried Australian ballet dancer Robert Helpmann, and old groveling obituaries in general, with these two memorable paragraphs:

His appearance was strange, haunting and rather frightening. There were, moreover, streaks in his character that made his impact upon a company dangerous as well as stimulating.

A homosexual of the proselytizing kind, he could turn young men on the borderline his way. He was also capable of cutting a person down in public without mercy. Yet many will remember him as an amusing companion. . . .

The *Daily Telegraph* immediately began making memorable mischief with obituaries in response. Two years later, the *Guardian* leapt into the fray. The result was nothing short of a revolution, the Obituary Revolution, which sent shock waves through the English-speaking world and created a generation of fans. The revolution was in the form of the obituary as the *Independent* fashioned it, and in the anything-goes style of it as the *Daily Telegraph* practiced it—but it was also in the relish with which these papers and the *Guardian* and *The Times* pursued their subjects. Nearly twenty years later, London obits still dominate, in quality and quantity.

The closest to this scene in the U.S. is in New York, but only the *New York Times* and the *Wall Street Journal* can pass for national papers, and the *Journal* doesn't run obits, only lists of the dead. Every one of the four papers I picked up in London considers itself a national newspaper, covering the whole of the British Isles and the world beyond. The success of their obituaries is tied up in that hovering ambition. They don't just obituarize the outstanding citizens of the United Kingdom and the big world figures. They cover Americans more thoroughly than the Americans do—it's not surprising to read a long and fascinating obit of a Mississippi blues man or a Hollywood producer or the Yank who bombed Nagasaki wedged between a Scottish singer and an Oxford don. The writers and editors leave the messy sentiment of local obits on the cutting-room floor; the few paid obits that appear are neatly corralled. You notice only the glorious faces, charged with significance and deliberately elevated, laid out with care and just the right amount of white space. The closest thing we have on this continent is Canada's *Globe and Mail*. In the United States, only the *Los Angeles Times* approaches the sheer aesthetic beauty of the British obits page.

How the British obits look is part of their impact, but how they read is the point. A great British obit doesn't read like a prosaic résumé. It's an opinionated gem of a biography, informed by all kinds of history, high and low, including gossip. It has the clear-eyed perspective of an op-ed piece and the drama of the news. It doesn't pull its punches in consideration of the dead; it aims not just for factual truth, as one of the editors put it, but for some sort of "higher truth"—and it takes pleasure in its aiming. Most of all, it is, in its individual state, a highly particular and layered creation, and something of an acquired taste. (You might, like me, want to pass on the cricketers.) The only way to demonstrate is to plunge into them, and mark up those pages.

My first London deaths, spread out on the bed, the floor, and the dresser, were modest. Arthur Miller had died the week before, the pope would live another six weeks, and Camilla Parker-Bowles was still breathing, preparing for the ever-diminishing spectacle of marrying the Prince of Wales. But no day is a disappointment in London, and this day gave me two great subjects to compare, a future saint, who as a little girl saw the Virgin Mary in a tree at Fatima, and a political martyr, the former Lebanese prime minister, who'd been assassinated.

The *Daily Telegraph* with its big broadsheet pages looked impressive, and completely covered the bed. The *Telegraph* is conservative and brash and flamboyant, the print equivalent of its flamboyant obits editor, Andrew McKie, who stormed through the obits conference. On a big page that included a mock speech by Tony Blair, the editors ran the assassinated former prime minister Rafik al-Hariri below the fold with a laughing photo of him receiving election results, his "beetling eyebrows" looking drawn in with a Sharpie. The billionaire had been known as "Mr. Lebanon" for his success reconstructing the Lebanese economy. "When it was once suggested to him that the Lebanese economy would collapse if he died, he replied, 'So, keep me alive.'" In and out of the prime minister's office five times, he had just criticized the Syrian occupation of Lebanon the day before he was murdered. Had Syria done it? The *Telegraph* didn't speculate, and at any rate, the miracle nun, Sister Lucia (full name: Sister Lucia de Jesus dos Santos) was judged more important. Her obit dominated the *Telegraph*, with a three-column spread of the 1917 picture of three frowning little Portuguese shepherd children against a stone wall, the two girls in nunlike veils, and the boy between them wearing a kind of jester's stocking cap. It's an eerie old picture, a great, weird thing to come across on an obits page, and the story was perfect, with its ghostly spirits, the great flu pandemic (which killed two of the three Fatima children), and the third child (our obit subject) fleeing like a

Hollywood celebrity to the refuge of a cloister. When the Virgin Mary first showed up in the rocky pastures to tell the children to pray the rosary—and also to tell them the first World War was about to end, and a second would begin if "humanity did not repent"—the children were assumed to be liars, the apparition their way of crying wolf. A Portuguese official, the *Telegraph* claimed, "kidnapped the children and . . . threatened to boil them alive in oil if they did not deny the allegations." Yes, boil them in oil—you know it's the *Telegraph* by those over-the-top details. "When this failed, he cast them for a night into the county jail, where [one of the children] led the prisoners in prayer and [another] danced the fandango with a thief."

Never mind that there was no one alive to verify the actual dance this child had done, much less the type of criminal she was doing it with—could any of these papers top that "danced the fandango with a thief"? The liberal *Independent* doesn't have the blithe and reckless spirit of the *Telegraph*, but some aficionados give it higher marks for its thorough, stylish, and insightful obits. The *Independent* recently turned tabloid in a cost-cutting move; this has the advantage of making it easier to read in a closet-sized room. Its lead obit page didn't look very beautiful that day, the type broken by somber/scary headshots of the miracle nun, swaddled in particularly constricting nun garments, with thick glasses askew, and Hariri, looking even more strangely beetle-browed in a formal portrait. But it was a great page anyway, because its third obit—of a missionary nun stationed in the Brazilian rain forest who had been murdered because of her protests against clear-cutting the forests—made an elegant theme of nuns and political martyrs (to say nothing of confirming my long-standing theory that people frequently depart this world in matched sets). And all three obits on the page were deep condensed biography, fascinating current history. There was even a joke: at the end of Sister Lucia's obit, the writer for the *Independent* noted that one of the ninety-seven-year-old nun's last visitors in the cloister was Mel Gibson, who, in spite of the fact that the recipient was deaf and blind, presented her with a DVD of *The Passion of the Christ*. Like they had a DVD player in the cloisters!

Left-wing, and solid as the schoolteachers who read it, the *Guardian* casts a cool eye over the political and cultural landscape. The handsome broadsheet, with its big, gorgeous photographs, ran Rafik al-Hariri, "Rags to riches Lebanese premier" top left, and artist Paul Rebeyrolle, who'd died a week earlier, next to him, looking very much alive and roguish in his studio, cool modern paintings propped in the background. The *Guardian* saved the Fatima nun till the next day, when it could run a giant black-and-white photograph of the three frowning little Portugese shepherds.

The Times, more than two hundred years old and owned by Rupert Murdoch, is also now tabloid-sized. It devoted the most space to the obits, running five in two and a half pages, but Sister Lucia and Hariri weren't among them; only one of the obits was newsworthy, a French race-car driver who streaked across the second page in a gleaming red Ferrarri, and had died just a few days before. The subject of another of its obits had died six weeks earlier. There are days this newspaper nails its obituaries, flashing wit as it writes the first draft of history, but this wasn't one of them.

If I were home in New York, I would have read all these online and missed the visceral pleasures of blackening my fingers, seeing the pages in full stretch with their telling photos and varied typefaces while I scrawled underlines and exclamation points and circles around the text. The Internet had brought me to these newspapers, but aside from its admittedly essential world-shrinking function, the computer seems a poor medium for reading obits. The inherent impatience of the machine, its constant hum, its cursor beating even at rest, the ads glaring along the borders—all conspire to speed you up. And there are so many interesting people dying, and the cascade of the dead is so relentless, that speed-reading obituaries can cause you to miss the point. Obit reading is an act of contemplation. I needed to hold the British obits in my hands. Obviously, I needed to fly to London, fill my arms with papers, and lie around a rented room for hours, wallowing in the physical pleasures of London's obituaries.

Only after such a banquet of newsprint was I ready to hit the streets. I wandered through Piccadilly Circus and Trafalgar Square, squeezed past the red double-decker buses and shiny black cabs, and ducked into a nearly three-hundred-year-old church, St. Martin-in-the-Fields, where an angelic harpist was plucking heavenly music at a lunchtime concert—a fitting epilogue to my morning. I followed the crowd out of the church, slipping around to a side entrance to the crypt, a musty basement with arched ceilings. The church runs a café in this crypt. I went through its cafeteria line in the semi-darkness as bare-armed lunchroom workers lifted vats of steaming comfort food, and filled my tray with cauliflower soup and bread-and-butter pudding. I ate at one of the little tables, and rested my feet on a gravestone from the 1700s, re-claimed from an old graveyard and cemented into the stone floor.

I didn't need a guidebook; everything in London spoke to the obits. A short walk away was Westminster Abbey, which held the graves of the great kings and queens of England, and Darwin's bones, and Shelley's ashes, and hundreds of other moldering ancients, not far from the Imperial War Museum and the warren of rooms where Churchill bunkered during the Blitz, where one could vividly imagine the bombs exploding into all this history.

Could there be a better destination for the obituary tourist?

BOILED IN OIL, AND OTHER
TERRIBLE FATES IN THE *DAILY TELEGRAPH*

Black cowboy hat on his head, a pocket watch chain dangling from the vest of his dark suit, Andrew McKie guided me across the windy plaza in front of the *Daily Telegraph*'s high-rise offices in Wapping. I was gushing about my four-newspaper morning. "What? Only four?" he said. He reads six a day carefully, and looks at ten. That's one benefit to his commute, a train from the suburbs and then another half-hour out to the Canary Wharf/Docklands business parks.

Once upon a time, the *Telegraph* had been firmly ensconced in Fleet Street, nestled near bars named the Falstaff or the Red Lion—and one nicknamed the Stab in the Back—frequented by tough-guy reporters. Those scrappy days ended in 1986, when Rupert Murdoch led the exodus to

Wapping. This area had been bombed in the Blitz. Remade with glass and steel, planned, updated, and computerized, these offices were no bleaker than a corporate park in the United States. There had been a few ingenious accommodations to the sort of workers who found themselves out there—for instance, between the tube station and one of the few buildings you might call a high-rise was a concrete plaza with a spacious, smoky franchise bar, the Slug and Lettuce, our destination. I expected the Slug and Lettuce to be slimy, but I found only well-dressed, cheerful clusters of office folk drinking their beer, urged on by driving pop music. McKie cocked an elbow on the bar and a boot on the rail and considered. He had given up liquor for Lent, and had been without libation for a whole week, but it took him less than thirty seconds to decide this was an extra-Lenten occasion. We got big mugs of Guinness stout and spread out my clips and a few days' *Telegraph*s on one of the well-lit tables.

The *Telegraph* had recently run the obituary of an outrageous British comedian, Malcolm Hardee—an obituary that did him justice—and we savored it all over again while McKie rolled a cigarette. The *Telegraph*'s obits run without bylines, but McKie claimed credit. It sounded like him. Hardee, "a notoriously dangerous sailor," had drowned in the Thames, but while alive, he'd made great copy, and McKie had a jolly time recapping his exploits.

> *He did an impression of Charles de Gaulle, his penis playing the part of the General's nose. He was also celebrated for a bizarre juggling act performed in the dark and with nothing visible apart from his genitals, daubed with fluorescent paint.*

Fans would greet his arrival on stage with cries of "Get yer knob out." He was said to be huge in Germany and Sweden.

"My favorite line," McKie pointed with satisfaction, "'He was said to be huge . . .'"

McKie was, at that moment, on deadline. I had been late, wandering through the tubes underground, and he had an imminent meeting at which it was entirely possible his new editor would announce substantial editorial layoffs. He had also been up half the night with one of his daughters, who was two and had become "violently sick" all over him, not once, but twice. Was he flustered, rushed, irritable? Not a bit. With his great rubbery comedian's face and his tumble of words, and the smoke curling up around his head, he looked devilish. Miss an opportunity to expound on obituaries? Noooo way.

I asked him about the scowling Fatima children, and the suspicious business about boiling them in oil. "I don't think the district regulator *would* have boiled the children in oil," McKie said thoughtfully, "but remember this was 1916, Portugal was very, very backwards, as was Greece until very recently, and Spain really only becomes a country under Franco, whatever you think of Franco—and there's some question over whether Italy is really a country even now!" He squinted through the smoke, perhaps to see if we would stop to have an argument, but I didn't want to waste time talking about fascism when there were obits to discuss; so McKie barreled on. "He may have thought [threatening to boil them] is just the way you get information out of children. And it's a good story!" The obit had come from a regular contributor who filed it years ago, but she was reliable, McKie said; "she

read all the memoirs." Sister Lucia's death had been the occasion to run that great picture of the Fatima shepherds. "This
was one of the very rare occasions when you can use a picture
of somebody as a child." He studied the unfortunate nun picture the *Independent* had run. "Not very bonnie, was she?"

Hariri had given McKie a spin. He had died before noon,
and McKie couldn't reach his Middle East expert, and had no
assurance he would reach him—maybe the man was on vacation?—so he'd rolled up his sleeves and started writing it himself. He was "absolutely astounded" that *The Times* had failed
to run an obit of Hariri. "That man was obviously of quite serious strategic importance to the Middle East. This is a big
story, Syria and Lebanon, one has to assume the Syrians have
done it, though that's for the news pages to speculate, but it
certainly has all sorts of political implications, and—it's a billionaire story! Now this is somebody who is going to be all
over your news pages like a rash tomorrow morning. I'm
thinking you're going to look daft if you don't run him. If he's
died by noon, I really think I ought to get a bollocking by my
editor if I don't get it in." His Middle East expert, as it
turned out, called back and filed in time, saving McKie a bit
of sweat, but he would have gotten it in regardless. He
flipped through *The Times*'s obits dismissively. "We did
him . . . we did him the day before . . . we did that one," he
said, a kid flipping through baseball cards: *got it, got it, got it.*
He paused. "We didn't do that one, but we have *him* ready to
go, and . . . this one, I noticed him about six at night and
thought it would look smart to get him in, but there was no
one to write it."

All in a day's work. McKie figured that he and his deputy at the *Telegraph* and a third staffer write half of all the obits, and farm out the rest, "but everything that we farm out gets rewritten. We have experts, some of whom write very well and entirely in our style, and it's just a matter of making them fit. But we have some people who can't write for toffee but know their stuff, and frankly, the info is more valuable than the writing." He claims that he can bat out 1,500 words in forty-five minutes if he has a good stack of notes or old news stories. "Partly that's helped by the template. The template is your friend! Once you know the rules, you can break them happily when you know why you're breaking them, and what you're breaking them for."

Variations in the standard ending ("She leaves a husband and six children"; "He was unmarried") sometimes come about because there's a stray fact that the writer hasn't been able to jam in elsewhere. "'He never learned to drive' was left over once and somebody said, 'Let's put that in as the last sentence! It would be funny!'" McKie chortled. "So we did. It turned out—misinformed! Not only had he learned to drive, he'd driven every day of his life. So this became a running joke in the office. I often use it as a last sentence.

"The way to do someone ludicrous is absolutely straight," he declared. If it's, say, "the producer from Run-DMC or Aaliyah, you should write their obituaries the same way as you would write a particle physicist. You've got to talk about these people as if you were talking about Watson and Crick. That's what makes it funny." He defended himself against critics who think the *Telegraph* sneers. "Sometimes all you're doing is putting a bit of pizzazz into something that's actually quite flat. You might have somebody who's very worthy and nice and good, but it's slightly dull. And you ask the family, 'What was he like?' You fish desperately—was he good at Scrabble, were there any foods he hated, did he spend all his time playing golf? And I remember doing this once, and they said—he couldn't abide ratatouille! And I put that in for my last line. 'He couldn't abide ratatouille or pesto.' I'm proud of that!"

He had left an obit in progress on his desktop to meet with me. "I have written my first sentence: 'Owen Allred, who died on St. Valentine's Day, aged ninety-one, became Presiding Elder of the Apostolic United Brethren, a schismatic branch of the Mormons which advocates polygamy, after his brother Rulon was shot dead in 1976 by the thirteenth wife of Ervil LeBaron, leader of the Church of the Lamb of God.' You want to read on, don't you?" Yes, I told him, every one of those details left me slavering. He was pleased. "Now, frankly, after that, it doesn't much matter what I write. But at the same time, I'm not going to go sneering at polygamous Mormons. I'm religious myself, I wouldn't sneer at them. I'm an Anglo-Catholic, a member of the more catholic wing of the Church of England—a high Episcopalian in America."

I looked up his Owen Allred obit later. McKie had indeed written it poker-faced, right up to the end. After mentioning that Allred was survived by eight wives, twenty-three children, and more than two hundred grandchildren, McKie referred to one of his wives as "his better eighth."

I also hunted down his obit of the ratatouille-hater (www.telegraph.co.uk, to "obituaries," then I typed in the search box "ratatouille"—George Guest popped right up).

That had been a memorable obit. Guest, a choirmaster and organist, had taken his choir to Australia and returned "in an aeroplane bristling with 'aborigine spears and those things that come back.'"

McKie frequently practices the writers' art of making lemonade out of lemons. When he ran one obituary prematurely, he followed it with an essay-length correction called "The Day I Managed to Kill Off Tex Ritter's Wife." It concluded:

> *I apologise unreservedly to our readers for having misled them. More importantly, I apologise to Mrs Ritter. I am genuinely delighted she is still with us—I came to like her a lot while preparing her obituary for the page.*
>
> *She may even have the good luck to follow Cockie Hoogterp, whose premature obituary The Daily Telegraph published in 1938. After 50 years, during which she sent back all her bills with the word "Deceased" scrawled across them, it was referred to again in the newspaper. She then wrote in to say "Mrs Hoogterp wishes it to be known that she has not yet been screwed into her coffin."*

The cost-cutting meeting loomed and McKie was drinking a second stout in preparation. "There are ninety of us being sacked," he said, out of about 500. "It could happen anytime in the next three months." It seemed inconceivable that McKie could be fired. Where else could a publication find such an energetic and cheerful obituarist to carry on the work of Hugh Massingberd, the father of the *Telegraph* obit? "Well," McKie said brightly, "the focus groups tell us we are one of the most popular items in the paper."

Massingberd still hovered over the *Telegraph*, an eccentric who spoke the language of Evelyn Waugh, P. G. Wodehouse ("The Master," they call him at the *Telegraph*), and the seventeenth-century gossip John Aubrey, whose catty pieces about Shakespeare and Milton, collected in his *Brief Lives*, still entertain. With these wits as inspiration, Massingberd had kickstarted some life into the *Telegraph*'s obit page back in the magical days of 1986. For seven years, he and his deputies ran riot, until he was carried out on a stretcher, the victim of a massive heart attack. No more deadlines for him, though he did edit a handful of the *Telegraph*'s collections, each dedicated to an obit editor who succeeded him in carrying the flag. "I have a vision," Massingberd had written, "of all the 'illustrious obscure' figures from the Raj, the Empire and the Services, the legions of dotty dowagers and sterling squires from, as one of our old school songs put it, 'the great days in the distance enchanted' . . . who could adorn the obituaries page of the *Telegraph*."

A FEW WORDS ABOUT THE CODE

It might be useful to step back for a moment and consider another contribution Hugh Massingberd made to the art of obit writing. He was a master of the euphemism.

In the folksier regions of America, where these euphemisms abound, they begin at the very top of the obituary. It's considered too bald and mean to say outright that somebody has *died*. People *pass*, or they *pass on*, or something even more strange and colorful takes off with their spirit. *He joined the choir eternal. She's knocking on heaven's door. He's gone to the rainbow. She went to paint the Pearly Gates.* These euphemistic ribbons dress up the dreary business of death, and give writers an opportunity to ply their trade. *She was promoted to Glory. He earned the golden halo.* Or, as the *Houston Chronicle* had it, *She accidentally went to Jesus.* My mother called me the other day, terrifically excited: Someone in Little Rock had just *left to play accordion in Jesus' band of angels.* Even readers who believe in a fundamental heaven with a literal choir know the writers really meant *he died.*

A newspaper is the biggest dose of reality that regularly comes through our doors, but it isn't like the raw satellite feed that occasionally shows up inadvertently on your television, with its glimpse of a snarling anchor or the first lady hissing cues to her president. Its contents have been groomed and edited. Newspaper quotes, for instance, aren't literal transcripts of what someone said; they've been cleaned up and pruned of their, *you know*, meaningless phrases. Most of the repetition, the bad taste and bad grammar, and all the obscenity, has been neatly sliced out. What's left has been vetted by the lawyers (the *killers* turned into *alleged killers*, etc.) and squeezed into a journalistic mold. A newspaper is a dose of reality that's been processed.

Most of us understand this. We figure that the *alleged killer* probably *is* the killer. When the obit says someone was *vivacious* or *sociable*, we can guess what's behind these chirpy, mom-approved words: the departed liked to party; she felt comfortable on the barstool. Jude Law, playing an obit writer in *Closer*, got points with Natalie Portman and his audience by translating the code used at his newspaper: 'He was a convivial fellow,' meaning—he was an alcoholic. 'He valued his privacy'—gay. 'He *enjoyed* his privacy'—raging queen!'"

This coded understatement is an art, and part of the pleasure of reading and writing obits. Massingberd spent his career at the *Daily Telegraph* refining that art. "We all know 'he didn't suffer fools gladly' translates as 'a complete bastard,'" he told a gathering of obituarists in Bath, England. Massingberd is a great elegant bear of a man, and in a self-penned mock obit claimed to possess "an appetite of such magnitude that friends counted him three men at their table." He smacked his lips over his list as if it were a tower of profiteroles, then read it with lusty pleasure:

Gave colorful accounts of his exploits—Liar!
No discernible enthusiasm for civil rights—Nazi!
Powerful negotiator—Bully!
Tireless raconteur—Crashing bore!
Relished the cadences of the English language—Old windbag!
Affable and hospitable at every hour—Chronic alcoholic!
He was attached to his theories and sometimes urged them too
*   strongly—Religious fanatic!*
Fun-loving and flirtatious—Nymphomaniac!
An uncompromisingly direct ladies' man—Flasher and rapist!

Phrases like these romp through the dozen or so collections of obituaries that the *Daily Telegraph* has published, mocking eulogies and puncturing any atmosphere of hushed respect. The Reverend Peter Gamble's "gifts as a teacher were considerable, as was his appreciation of beautiful boys." The *Sunday Express* lost two million readers under Sir John

Junor's watch, and so, perhaps "There is room for debate about Junor's success as an editor." Bapsy Marchioness of Winchester, who pursued her philandering husband across continents and stopped only to send out press releases, was called an "enthusiastic self-publicist." The *Telegraph* never uses the words "pederast," "disastrous failure," or "raving mad exhibitionist." It doesn't have to. The newspaper employs understatement and mock-delicacy, not to avoid saying something baldly, but to set up the joke.

> *Miss [Hermione] Gingold had an endearingly individual approach to life. In New York she was regularly seen rummaging through other people's dustbins.*

> *Not known for his delicacy towards the fairer sex, [Peter] Langan rejoiced in daring attractive young women to strip naked in the bar in return for limitless champagne.*

The reader who takes delight in the code runs the risk of seeing it everywhere, even where it isn't. Take the phrase "She seemed to thrive on mystery," which appeared recently in an obituary. What does that mean now—she was a conniving, double-crossing, femme fatale? Maybe she just liked to wear scarves and veils. "He was a man of boundless and sometimes impish energy." Is this double-talk, and if so, how do we translate it? He was annoyingly elflike? He took methamphetamines? As Massingberd himself admitted, "'He was unmarried' could mean anything from, well, 'he was unmarried,' to 'a lifetime spent cruising the public lavatories of the free world.'"

These days, there's as much blunt talk in the obits as there is delicate obfuscation; euphemisms are as quaint as those funeral parlor fans that women once used to beat the air during a hot service. It's okay to mention AIDS and cancer, and it's okay to accuse a whole culture of being ridiculous.

> *Dr. [Alain] Bombard became an instant legend in France in 1952 when he drifted from the Canary Islands to Barbados in a small rubber boat. He joined a long list of Frenchmen who have performed seemingly silly feats at great hardship and, often, immense risk.*
>
> (*by Douglas Martin,* New York Times)

Even the sentimental obituarists with their palette of stock phrases—*friend to all, died doing what he loved, would give you the shirt off his back*—slip some tough reality into the mix. As for the writers who consider tough reality an essential element in their obits or life stories, the facts are the facts, and neither readers nor survivors are coddled.

> *Robert Davolt, a San Francisco leather luminary who immersed himself in the world of sadomasochism for more than two decades, has died of melanoma at the age of 46. . . .*
>
> *A celebration of his life will take place at Daddy's Bar in the Castro on Saturday afternoon. . . .*
>
> *Mr. Davolt was the last editor and publisher of Drummer magazine, a leather journal that closed in 1999. He staged leather contests, wrote a book titled "Painfully Obvious: An Irreverent & Unauthorized Manual for Leather/SM," served as editor of male bondage magazine Bound & Gagged until he*

got sick, and did an online column for leatherpage.com and its 125,000 readers. . . .

(*by Patricia Yollin*, San Francisco Chronicle)

The sad fallout of a woman's life is described bluntly.

"I could write a book," [Dorothy Custer] said often after rearing 13 children.

Her daughters encouraged her to do just that. But it was too late. Launching and overseeing so many young lives took her own well-being. When later she didn't need to be "Mom" anymore, she had no energy, will or nerve. Even before her last child was born, she had begun using alcohol to retreat from her burden, and possibly, from depression. Social drinking became wine every day at 5 P.M., and then, all day. Intervention was not successful.

(*by Amy Martinez Starke*, Oregonian)

Why not celebrate the leather luminary's real life? He did. Why pretend that thirteen children is the stuff of a sitcom when it obviously used this woman up? She knew it.

My favorite—and the clearest evidence that the code has become an optional tool, best used in conjunction with a wink—is this sentence from a tribute to one of the men who started the CIA:

[Walter] Pforzheimer had a private income (his father and uncles made fortunes in Standard Oil stock), never married and was often described as crusty, outspoken and a curmudgeon—euphemisms for his conservative politics, social views

that included crude prejudice against Jews and blacks and a
manner that could veer from fawning on the great to public
abuse of menials.

(*by Thomas Powers*, New York Times Magazine)

I think when you call a euphemism a euphemism, it isn't a
euphemism anymore. But as long as there is a *Daily Telegraph*,
this convention will have a shrine.

FOLLOWING THE *GUARDIAN* INTO THE MIST

One of the *Guardian*'s obits editors, Nigel Fountain, had been
periodically calling the rowdy Groucho Club, where my husband and I were staying in London, with reports of his
progress through the British Isles. He was on a car trip, a
week's vacation from the obits, researching a book about old
theaters named the Empire (there are apparently dozens,
some in ruins). He had logged hundreds of miles by week's
end, but he insisted he wasn't too fatigued to escort a couple
of foreigners to dinner. He called three times on the way to
London to report his looming approach, the last to announce
he'd pulled up across the street. He swept his bags and papers aside and declared he was springing us out of the West
End. It was a Friday, late winter, at twilight. The streets of
Soho were teeming already with office workers, theater-
lovers, tourists, druggies, pimps, pushers, thieves and journalists. Fountain dodged them expertly, and pointed the VW
north.

He was the only source I hadn't reached through his publication. Somebody had met someone over the Internet who insisted this man Fountain would take care of us. Was he even employed by the *Guardian*? Was this the beginning of a cozy British murder mystery—the obituarist luring the New Yorkers into his car, taking us up to the Heath to dispose of our bodies, and then running back to the office to write us up? I had gotten the impression, via his phone calls, that he viewed obituary writing as a desperate activity.

The narrow streets and canyons of London gave way to lower-slung buildings and more open boulevards. He pulled off into the mists of Kentish Town, then led us behind what seemed to be a motorcycle repair shop into a semi-deserted bar, through three or four darkish rooms into a garden of picnic tables lit by tiki torches. He would feed us, apparently, before killing us.

The food was delicious. I remember buckets on our picnic table—mussels, bottle after bottle of good red wine, giant bread baskets—and a shadowy figure who kept bringing plates. Fountain, aged sixty, who had the casual slouch and the patter of one's favorite professor, entertained us with stories of his days on a series of radical papers, bumbling into the *Guardian*, writing his first obit of Abbie Hoffman, filling in for a friend on the obits desk in '94 and then sticking around to assign and edit for the obits editor, Phil Osborne, occasionally in a frenzy. When the poet laureate Ted Hughes died, Fountain was on duty and found that the writer assigned to write Hughes's obit in advance had never bothered. Fountain tracked down a poet in Oxford who agreed to write one instantly. We were in suspense. Hughes, the ex-husband of Sylvia Plath, and a beloved poet himself, was a huge figure in the U.K.; excerpts of his work had been laid out to run in the front of the paper. Not having an obituary was unthinkable. "Come five-thirty, no trace. Come six-thirty, no trace, and creeping alarm because the page's first edition was due to go at eight. I spent my time ringing Oxford, where the writer was based. I rang his college, rang his landlady, spoke to a Chinese person in college middle common room—no trace of him anywhere. Mounting panic. Did the writer exist? Had he written it? Did I have a job? I start writing it myself, 'It was twenty years ago that a young Ted Hughes could have been found—' Phone rings. It is the innocent obituarist, calling from a phone box with the sound of Oxford bells in the background. Has the piece arrived? He sent it as an email. No, the piece had not arrived." Fountain ordered the obituarist back to his computer. "Minutes pass as hours, the screen remains angrily empty." The editor of the paper paces in front of them, until finally, with the intervention of a crack tech team, the text materializes, is whipped into shape and fitted to the space, and the obit of Ted Hughes squeaks into the first edition. Fountain and his assistant editor retire to a nearby pub and collapse.

"Obituary writing is regarded as a staid activity, when actually it is a *terrifying* activity," Fountain said with feeling. He had written Linda McCartney himself one evening, past deadline, knowing nothing—though he faked it so well *The Dictionary of National Biography* asked him to write her entry. "Lately I've been writing the crew of the *Enola Gay*," he said. The history interests him. In an obit he did on a crew member of the *Bock's Car*, which dropped "Fat Man" (nicknamed for Churchill) over Nagasaki, Fountain threw in an eyewitness account he'd picked up in the oral history archives of the Imperial War Museum.

"All these people are the history of the twentieth century," he said. "Brick by brick, we're building a wall." Unless he said we were "building a mall"—my notes, written in the dark on a rough surface and splattered with frightening red blotches, like a Ralph Steadman painting, are hard to reconstruct.

Fountain believes the Obit Revolution of 1986 was kicked off by technology, when the newspaper business shifted from printing presses to computers. "This led to the birth of the *Independent*, which made a big deal of its obits, and ran them with bylines, which opened up the idea of a more opinionated line of obits. The other three heavy papers were forced to respond." By 1988, the *Guardian* was off and running with its own smart and culturally savvy obits, also bylined. The *Guardian*'s circulation is about 400,000 (the *Telegraph*'s is over 900,000, *The Times*'s just under 700,000), but the *Guardian* has ambition (within months, it would launch a re-designed obits page) and stringers everywhere. Fountain said it has a stringer in Italy, just to keep tabs on deaths in the Italian cinema. It takes pride in covering popular culture. I couldn't find any other newspaper that ran an obit of Martha Carson, the country-and-western singer whose "trademark stance of dropping to one knee at moments of emotional climax, the microphone stand at an angle" had been picked up by Elvis in the early fifties, though the *Washington Post* mentioned her in passing.

By then, we'd relaxed in the tiki garden, mellowed in the late-winter warmth as old ghosts—the departed Beatles and Beatles' wives and Lady Di—swirled around us. That would have been the time for a poisoned nightcap; we'd hardly have noticed. Instead, Fountain shepherded us to the VW Golf, and pointed south. But he felt badly that we hadn't seen more of the city, so he drove back circuitously, looping around London, pointing out landmarks in the dark. I have no idea what we saw, though it all looked lovely in silhouette. No, that's not right. One place stayed with me, a bulwark, right in the center of town near St. James's Park, an old fort, a solid wall of bricks hundreds of years old.

AN *INDEPENDENT* BENT

James Fergusson sat in Soho in a corner of the Groucho Club, which had emptied of noisy, drunken journalists a few hours earlier; it was alive in late morning only with kitchen staff, polishing glasses and vacuuming ashes. Fergusson didn't have time for an interview and was running more than an hour late that day, but in a rash moment he had promised to stop by the club on his way to the *Independent*'s offices in Wapping. Now he was faced with this obituary enthusiast (me), shaking out the page with the two nuns and the Lebanese prime minister, and I could tell, if there was one thing that bothered him, it was all this *obituary enthusiasm*. When he launched the new-style obituaries in the *Independent* almost two decades ago, he had done so with the intention of drawing in new readers. He hadn't anticipated these new readers would turn into rabid, campy fans who pursued him for interviews.

An impeccable man with an elegant gray suit, soft, curling

gray hair, and horn-rimmed glasses, James Fergusson staked out his contrary position almost before saying hello. "What I find disturbing is this sort of cultishness which seems to increase around obituary writing and reading," he said. He had cooperated with a magazine writer for *Vanity Fair* a decade ago, and was still traumatized. "They decided to have a photo shoot, and they would pick up all the obit editors and take us to Kensall Green Cemetery, and I refused absolutely. Hugh Massingberd and some of his staff attended, and they made them stand in a newly dug grave with a mist machine playing filmic mists in the background." He shuddered.

Fergusson had come to the obit page not as a journalist but as an antiquarian bookseller. He had been "a lazy classicist" at Oxford who stuck around after graduation and ended up running Waterfields, reputed to be the best bookstore in Oxford. Several of his more ambitious pals announced in 1986 that they were going to start a new national newspaper, to be called the *Independent*. It was one of those idealistic projects that aging lit majors think up: they would rethink all the conventions of the newspaper, reinvent all the forms. Fergusson had some suggestions for the books pages, and his friends set him up with Andreas Whittam Smith, the new editor of the *Independent*, but Smith already had a novelist working that job. Smith looked Fergusson over, though, and wondered what else he might do for his paper. "And I said, 'Do for your paper? I'm a bookseller. I'm very happy to write for you, but I deal in books.'" Smith called him back the next day. The *Independent* was thinking about adding an obituaries page; they thought he might be just the man for that job.

Fergusson "roared with laughter and put down the phone.

Then I thought, Well, all right, if you were doing an obituaries column from scratch, what would you do." And obviously what you would do would be signed obituaries," and for the young readers the *Independent* was trying to attract—"pictures and fun, and you would try and demystify them, because there are a lot of obituaries that are written in this strange code which you have to be a crossword solver to understand." By the time Fergusson had finished making notes and sweeping away the antique conventions of the form, he had projected himself into the enterprise, and talked himself into a new career. " I thought that we could make something of this, more entertaining, more transparent."

For Fergusson, the first shot in the Obit Revolution of 1986 was what happened when Rupert Murdoch busted the printers' union and took *The Times* to Wapping, beating a path out of Fleet Street: Suddenly there were wonderful *Times* writers quitting in disgust and looking for a new home. *The Independent* would never have succeeded, he felt, without all those great old *Times* writers. The new paper emerged at a crucial moment in newspaper history, and was an immediate journalistic success. In its nearly two decades, the *Independent* has struggled financially, been forced to give up its independence and suffered a succession of owners, but in spite of everything, it has survived. In 2004, it won a British Press Award for Newspaper of the Year. "Lots of downs and some jolly ups," Fergusson summed up its history. But unlike Hugh Massingberd, who lit a fire at the *Telegraph*, then passed the torch, Fergusson has been reinventing and honing and polishing his obituaries on deadline for nineteen years. If he had a showier personality, he'd be an institution.

The *Independent*'s obits, direct, often personal, and informed by a closely observed perspective, altered the form. The dreary lists of dates, affiliations, and survivors, his so-called *desperate chronology*, were broken out into a box, freeing the writer to tell stories and making the obit more of a loose biographical essay. "What Massingberd was doing at the *Telegraph* was just putting a twist on the old obituary," said Fergusson. "He was subverting the old obituary from within, whereas we were trying to change it altogether, from without." *Independent* obits can seem a bit windy and detailed if you aren't interested in the subject; but with a little patience, they can be the most rewarding.

Some composers produce music as a volcano spews lava or a tree sprouts apples, spontaneously, compulsively; others work like Faberge, meticulously constructing small-scale, perfectly designed works that display extraordinary craftsmanship. Arthur Berger was a jeweller, refining every detail in his scores, in an idiom that took what it needed from Stravinskian neo-classicism and Webernian serialism.

(*by Martin Anderson*, Independent)

The *Independent*'s obits are signed, and the first person often rears its head. It can be a shock to discover late in an obit that it's written by an acolyte, or former employee, or even the subject's sister or father. In an obit about a figure in the sixties music scene, you might come across: "I took Cream on their first trip to America. It was the most amazing week I have ever spent because they hated each other's guts and would spark one another off all the time." Newsman

Charles Glass sent off news anchor Peter Jennings with fascinating intimacy.

> *When I quit ABC in 1987 to write a book on the Middle*
> *East, I was kidnapped in Lebanon. It was the worst summer*
> *of my life and, coincidentally, of Peter's. When ABC broad-*
> *cast a tape of me confessing at gunpoint to being a spy,*
> *Peter—for almost the only time in his career—lost his profes-*
> *sional detachment and cried on air. His wife was leaving him*
> *that summer, and mine left me shortly after my escape in*
> *August.*

News purists may disagree, but I love these real voices ris-
ing from the page, assuming the authority to put lives into
context, but declaring their own passions and biases in the
process. And the *Independent* doesn't neglect the "japes," as
Fergusson called them, the wacky asides like the one in this
race-car driver's obit: "When he awoke from a 16-day coma
he found he could speak French, which was odd since he had
understood not a word of the language previously."

Fergusson said, "One of the great advantages of having
signed obituaries is that you can be a bit ruder in some ways,
because it's not the newspaper talking, it's the contributor
talking—but also, there is somebody answerable. There's
some sort of historical truth in what you're doing." He gives
the people who write obits plenty of leeway, letting, for instance, a publisher of Beckett complain in Susan Sontag's obit
about her "unauthorized" production in Sarajevo of *Waiting
for Godot*, or giving room to the writer of the Hunter S.
Thompson obit to complain about his influence on journalists

who write in the first person ("For that, he has a lot to answer for"). In an obituary of Kate Peyton, the BBC journalist killed in Somalia, her colleague and friend Fergal Keane erupted, "For her to be murdered in Africa is the most bitter injustice imaginable." Not one of the most bitter injustices, *the* most bitter injustice—the most bitter *imaginable*. Fergusson doesn't cut this excess; he runs it—he runs *with* it—and those extravagant flares of voice are one of the elements that make his obits page alive. Many of his obit writers have never written an obituary before, so he edits them for style and presentation, but he doesn't tinker with their voices or opinions. "Editing is to some extent, or should be, an act of empathy," he declared.

Fergusson and his "tiny, tiny staff," a deputy and perhaps one other part-time person, produce three to five obituaries per day, as usual, starting from ignorance and nothing but blank pages and a deadline. "Very few of our obits are written in advance, and if they are, they're out of date." And because of his high standards and his desire to get away from the lockstep format of "the desperate chronology," he can't simply fill in wire-service obits or dashed-off tributes. He wants each written by an expert in the field, not a generalist or a journalist or that odious invention, an obituarist. "People write me and say, 'I long to be an obituarist,'" he said, making a face. "But I don't want somebody who will just write and mug things up, because that's how journalists get things wrong. I want people to come from a particular speciality and have knowledge." What this translated to was chaos, every day. "You find out someone has died, and you ring up somebody who knows a bit more than you do, who says why don't you ring up someone who knows a bit more than I do, and eventually, you are confronted by a total stranger, and you don't know if he can write anything, and you don't know anything about the subject either. So a lot of guessing and instinct goes on and sometimes it works and sometimes it doesn't."

Sometimes it works very well, indeed, and at its best, it transcends biography. Fergusson ran an obit by "the brilliant cooking writer" Jane Grigson about a potato farmer who grew 367 varieties of potatoes.

> *When people rang to place an order, [Donald MacLean] talked to them for a while to make sure they were suitable—a character test, as if he were placing children for adoption. He felt that if people learned what quality was in one small sphere of existence, they would then be able to recognize quality in everything else—television, books, politics, social attitudes. Such aims were behind his endless parcelling of tiny orders. Salvation by potato.*

There's perfection in that slice of an obit, a crisp and simple truth in a tidy jacket.

As Aubrey's irreverent *Brief Lives* inspired the *Telegraph* obits, Fergusson shaped the *Independent*'s after the detailed and colorful obits in the great eighteenth- and nineteenth-century periodical, the *Gentlemen's Magazine*. "When I was at school, when I was bored, I would reach down for a volume. They were terrific—you didn't know what you would read next, which is the exciting thing about the obits page. It could be about anything, and all sorts of subject areas that people never read about in the newspapers. You can write learned disquisitions through the life of one person. It's great plowmen, or champion bagpipers—the great tapestry of human life. It's wonderful, that opportunity! If all one is doing is turning out another actress efficiently, then it's a rather boring job." He prefers to hunt down those rare birds, "the historian of typography" or "the cultural anthropologist who documented a hundred styles of sari draping."

"The more obituaries the merrier," he said. "We could produce twice as many pages every day than we do at the moment. You just try to fill the paper today with the best stuff you've got and throw the rest of it away."

The *Independent* is down to 250,000 readers, and cost-cutting has taken its toll. Fergusson has no top editors to back him up, no layout artists to guide his page or copyeditors to scan for errors, no fact checkers to verify information. The occasional picture editor will acquire a photo for him, but everything else is on his shoulders. "We have no cuttings library. We have no library at all. The largest collection of books is probably in the filing cabinets behind my desk. That the *Independent* does not have a library is shameful." That it produces quality obits day after day without one is heroic.

Fergusson found time to write an essay on obits, for a 1999 book called *Secrets of the Press*. In it, he pointed with pride to "the multitude who might never have had obituaries written about them if the *Independent* . . . had not come along," the people whose contributions and innovations fall outside the usual scope of a newspaper—"photographers, monks, bookplate designers, chairmakers, suffragettes . . . ," and, of course, potato growers. These obits of unsung people were British literary variations on Jim Nicholson's folksy homemakers and regular Joes, fit to a jazzier template, and mixed in with princesses and prime ministers. "The wide range of obituary coverage across British newspapers is its great glory," Fergusson wrote. "Here is rich material both for the reader . . . and for the historian: this is what Raphael Samuel, that ardent devotee of the obituarists' art, called 'the undergrowth of history,' thick with evidence."

Did I know Raphael Samuel's work? Fergusson wondered. Samuel had been a Marxist social historian at Oxford. In the mid-sixties he advanced the revolutionary idea that history didn't happen only to armies and kings, it also happened to ordinary people. His workshops on women's history, the history of childhood, and the culture of the immigrant had, as Samuel's obituarist for the *Independent* put it, "led people on journeys of creative self-discovery by blowing away the walls which separated working people from literary culture."

"Obits are one of the ways we can document that 'undergrowth of history,'" Fergusson said. And with this final, quiet nod to another revolutionary, he excused himself and faded into the London streets.

LIVES OF THE *TIMES*

That summer, hours after Londoners danced in the street celebrating the announcement that their city would host the 2012 Olympics, three subway lines and a bus exploded during the morning rush hour in the worst terrorist attack on British soil in history. "In 56 horrific minutes, familiar London landmarks became a monument to mass murder," *The Times* reported. It wasn't clear at first that this was the work of British Muslim extremists—boys and young men who also died in the blasts—or how many had been killed: 37, "approaching 80," 52, then 56. Was this more like 9/11, or the bombing of the commuter train in Madrid in 2004? "An Islamic website, claiming al-Qaeda responsibility, declared that 'Britain is burning with fear.' That was another lie, but anxiety and uncertainty, the terrorists' allies, gusted around the city," *The Times* wrote.

"Burning with fear" was exactly wrong. The Brits were steeled by anger, and determined not to be intimidated. They tracked the survivors in the press, following the progress of the girl in the burn mask pictured being led from one of the bomb sites; they were moved by the posters of "the missing" that cropped up almost immediately; and the whole nation came to a standstill for two minutes a week after the attacks to mourn in silence in an extraordinary display of unity and defiance. Then they briskly returned to work. They had been bombed during World War II, and as recently as the early nineties by the IRA. It was against their nature, and their experience, to burn with fear.

In fact, "amid the wailing of sirens just hours after [the] attacks," the founder of Amnesty International, Peter Benenson, who had died in February, was remembered in a memorial service in St. Martin-in-the-Fields. Benenson's friends, family, and supporters were visibly shocked by the tumult in the streets, according to the *Guardian*, but continued the service, "determined to pay tribute to his work."

Only one of those who died in the explosions merited a full

obituary in the *Independent*, Giles Hart, an organizer and long-time supporter of Solidarity and a culture maven who had been scheduled to deliver a talk that night on Lewis Carroll. "Our principle is to run things on the merit of subjects' lives, rather than based on the circumstances (however touching or historic) of their deaths," Fergusson wrote me.

But a week after the bombings, *The Times* began to run "London Lives" after its regular obituaries, profiling four or five of the victims each day. These obits were designed to be a seamless continuation of the regular obituaries. Where the regular obituaries ended with an abbreviated version of the *Independent*'s black box ("J. Neville Birdsall, New Testament scholar, was born on March 11, 1928. He died on July 1, 2005, aged 77"), the Lives began with these details ("Shahara Islam was born in Whitechapel, London, in 1985. She died in the Tavistock Square bus explosion on July 7, 2005, aged 20"). The text of each was about twice the length of a *New York Times* Portrait, and the photos ran grouped in the center of the page. The visual effect was elegant, dignified, and clear: these, too, belonged in the obituaries.

The text emphasized the homely details.

[Ciaran] Cassidy was popular in his neighbourhood and had a reputation for taking a long while to get anywhere because of his habit of stopping to talk to everyone along the way . . . Bars in Finsbury Park are displaying Cassidy's picture as a gesture of support to the family . . .

Described fondly as demanding, disorganized, unpunctual and "dotty," [Anat Rosenberg] was at the same time intelligent, loving, witty, supportive and loyal and obsessed with buying bags, jewellery and shoes.

London Lives pointedly described the religious affiliations and multicultural bloodlines of the predominately young citizens who had been caught that morning: the son of a Nigerian Catholic and a Muslim; a Vietnamese from Melbourne; the Polish dental technician.

> *[Mihaela Otto] lived with her close-knit, multifaith family in a detached house in Mill Hill East. She is survived by her 78-year-old Christian-Orthodox mother . . . her older sister . . . her Jewish brother-in-law . . .*

They were alert to the ironies, and provided a way to follow up on the dominant news story.

> *[Philip Russell] was among the crowds evacuated from the Underground at Euston station. He called his office at 9:30 am to say he was running late and would catch a bus. He caught the No 30 bus which was blown up in Tavistock Square.*

> *[Jamie] Gordon and his partner Yvonne Nash . . . had lived together in Enfield for several years. Nash's desperate search for Gordon in the aftermath of the attacks furnished one of the defining images of the tragedy, and was featured on the front page of* The Times *two days after the bombing.*

On some days they even appeared listed by subjects' names in the table of contents on the paper's second page. Most of all, they were, in their placement among the news obituaries and their integrated design, part of the larger message of how obituaries are changing. Even in *The Times*, even in the oldest and most establishment-conscious newspaper in the U.K., the young men and women who drank at the pub down the street and worked in offices and spent their money on bags and shoes had led valued lives.

In one generation,
a boring, moldy old form
has sprung to life.

Andrew McKie, *Daily Telegraph*; Dr. Nigel Starck, obituary
historian; and Claire Martin, *Denver Post*.

Charles Steck

"God is my assignment editor."

Richard Pearson, *Washington F*

"You take a fellow who looks like a goat, travels around with goats, eats with goats, lies down among goats, and smells like a goat, and it won't be long before people will be calling him the Goat Man."
Robert McG. Thomas, Jr., *New York Times*

Fred R. Conrad/New York Times

"I suffer and grieve thinking we might be overlooking someone deserving."
Chuck Strum, *New York Times*

"I organized this conference on a dare." Carolyn Gilbert, the Great Obituary Writers' International Conference

"Maybe there's a future in the past, I thought."
Adam Bernstein, *Washington Post*

"We know who's sick and who's dying, and we know how many of the original cast members of *Gilligan's Island* are still alive."
Amelia Rosner, alt.obituaries

"The way to do someone ludicrous is absolutely straight."

Andrew McKie, *Daily Telegraph*

John Farnham

"I think of obit writing as a way of
capturing souls."
Ann Wroe, *Economist*

Michael Freeman

"I have a vision of all the 'illustrious obscure' figures . . . who could adorn the obituaries page of the *Telegraph*."

Hugh Massingberd, *Daily Telegraph*

Linda Nylind

"Obituary writing is regarded as a staid activity, when actually it is a *terrifying* activity."
Nigel Fountain, *Guardian*

"A little life well lived is worth talking about."
Jim Nicholson,
Philadelphia Daily News

"I go out as if I have a camera on my shoulder. I almost produce my stories."
Jim Sheeler, *Rocky Mountain News*

"If people would take the paper for one thing, for the Life Stories, I'd feel like I've fulfilled my mission here."
Amy Martinez Starke, *Oregonian*

"He had a mojo bag given him by a voodoo priest." Catherine Dunphy, *Toronto Star*

"When someone dies [in Haines, Alaska] everyone helps—you know, brings a casserole, or offers to make the program, or play the piano at the service, whatever—and I can write." Heather Lende, *Chilkat Valley News*

"I'd be laughing and joking, falling out of my chair. 'What were you doing?' somebody would ask. I was doing an obit interview." Leon Taylor, *Philadelphia Daily News*

"If a few minutes of my own life, moments of irredeemable clarity [during the collapse of the Twin Towers] . . . are so difficult to get right, how much harder is it to present a truly accurate version of an entire life in twenty or thirty newspaper inches?"
Stephen Miller, *New York Sun*

10

Googling Death

When I began exploring the dead beat, I'd take a train into New York City and visit the main and mid-Manhattan branches of the public library to browse obits around the world. There I discovered, among others, the obits of the *Boston Globe*, the *Pittsburgh Post-Gazette*, the *Guardian*, and the *Globe and Mail*. Reading any of these papers, I felt as if I'd been transported to cities of substance and consequence, where people weren't hurried offstage after death but given generous farewells. It took a week or more for *The Times* of London and the *Guardian* to cross the Atlantic and for the Australian papers to make it around the world, days at least for the American papers; they were already history. But I loved to sit in the public library's great reading room, filling out old-fashioned requisition cards with a stubby pencil, then watching a librarian pull newsprint from stacks on the shelves, just like librarians of previous centuries.

By 2005, these papers and most others had easy-to-use websites; the entire archives of the *New York Times* had been scanned and were available online; and I could read *The Times* of London, and any other paper, in fact, the minute an editor posted it. I visit the reading room of the library now only for nostalgic purposes, or to study the physical layout of the pages with photographs. My computer has replaced the library, its big colorful flat screen cluttered with vivid icons, my desk swept free of messy clippings. The browser icon bounces when it needs attention; bells and whistles help me navigate the shoals of the Web. Most days, I don't use my legs, or the rest of my body, either—it's just my eyes and fingers, flashing impulses to my brain as fast as I can type, as fast as the screen can jump. Whenever my connection falters, I go mad.

Between the "Hi Marilyn Any Pills You Want" and the "word of the day," ("verjuice," the sour juice of unripe fruit, or acidity of disposition) was the email I'd been waiting for from Celebrity Death Beeper: "A celebrity death!" I'd signed up for the free service at deathbeeper.com to be instantly notified when someone famous died. The webmasters claimed to check the news blogs every ten minutes. I imagined getting a little shock when someone passed on, much as the children in *It's a Wonderful Life* heard a bell ring when someone got angel wings. I did get a tingle seeing "A celebrity death!" spelled out in my inbox. That jaunty exclamation point said, We're in the exciting business of passing along big, bad celebrity news. That sounded good. I wanted to be wired to the current that courses through a newsroom or ripples through the airwaves and cyberspace when someone famous died. I needed to know as soon as it happened, the minute it happened.

How deflating, then, when I discovered what celebrity deathbeeper.com considered a celebrity: Lygia Pape, seventy-five, "a multimedia artist and founding member of Brazil's vanguard Neoconcrete movement." May the survivors of Lygia Pape forgive me, but I'd been oversold. Could the deathbeeper have found a less celebrated celebrity? Is there a movement in any field, particularly an artistic one, with a more unfortunate name than "Neoconcrete?" Such a pedestrian beginning to my job monitoring the goings of this world! Beep me, I'd begged them. You want to be beeped? they said. We'll beep you!

I had a brief, self-loathing whirl with the website whos aliveandwhosdead.com and casketsonparade.com. I visited the deathclock, which calculated I had twenty-eight years to live (and where do I complain if I don't?). Some hours into one black hole of a day, I tumbled into findagrave.com ("Jim created the Find A Grave website in 1995 because he could not find an existing site that catered to his hobby of visiting the graves of famous people. He found that there are many thousands of folks around the world who share his interests"). *This is not it, this is not it at all.* I was wasting my time with graves and the celebrities who filled them. I didn't want the literal macabre; I wanted the metaphoric macabre. I wanted a central location where I could mingle with other obit fans, read obits, and not get stalked by creeps.

How do obituary fans find each other? By logging onto the public bulletin board known as Usenet. No matter how particular or weird our obsession, we can find like minds on Usenet. Go to google.com, click on *groups*, type in the phrase alt.obituaries. "Description: Notices of dead folks," it reads at the top. We have found the obit lovers' nest. It doesn't take long to figure out the basics of the alt.obituaries newsgroup. Down the page runs a list of the most recent posts—obituaries, mainly, but also political diatribes, jokes and health watches, alerts on hospitalizations that could bode imminent death. The subject headings read like obituary headlines: "Betty Jean Vallareale, Vegas dancer," or "James Nelson, killer who became a minister," or "Deuce, the Two-Faced Kitten"; or news headlines such as: "Veronica Lake's ashes found in antiques store." Reminiscences about friends and loved ones who have died appear next to accounts of the deaths of presidents. Prominent deaths appeared moments after being announced in the media. It's possible to watch an obituary develop—reading that someone has turned ninety, for instance, then seeing a notice of his hospitalization, a press release of his death, a wire-service obit, then a flurry of send-offs from various papers. The "official" obituaries are dissected, disputed, filled out; stories that couldn't be told are told. ("Almost everyone in Pittsburgh who loves baseball . . . loved Bob Prince," Bill Schenley, one of the core posters, wrote of a Pittsburgh Pirates broadcaster. "Unless, of course, they actually knew him. He was a miserable, mean-spirited drunk.")

Visitors to alt.obituaries might have to wade through some noise from the Libertarian or the guy still raving about Darwin to get to the good stuff, but it's worth it. This is the mother lode of obituaries, with public postings that anyone can read; and anyone who fills out a registration can post. The pace isn't the rapid conversation of instant messaging, but a stop-motion dance that feeds a steady infusion of records to a mysterious and seemingly bottomless library flickering out there in cyberspace. Its archive stretches back to late 1993, when the first alt.obituaries poster paid tribute to his dead cat. There were those who said it would never work. ("First of all, more than most groups this idea sounds like a joker magnet. . . . Who's going to stay with [it], week after week, wading through all these threads about people he's never heard of and doesn't care about . . . ?") In spite of such reservations, a decade later, the site racked up 56,000 postings for the year, most from people who had been posting to the group for years. That's a lot of talk about obituaries.

If the obit page of the daily newspaper is the floor of the Senate, or the official transcript, alt.obituaries is a smoky back room, the place where the deals are debated. Think of it as a clubhouse. The historian sits next to the guy who works in a tanker; the Upper East Side New Yorker hangs with the guy from the trailer park; obit writers mingle with people who have files bulging with clips. The walls of this clubhouse are scrawled with graffiti. In one corner someone counts the recently deceased poets, in another, all the famous people turning ninety that week. Members vie to be the first to post a new celebrity death. Occasionally one will take the floor to deliver a personal memorial. In the middle of this swampy mix of literary transcendence and fresh mud, one turns her head to whisper, "You can't make this stuff up."

I began to visit alt.obituaries after I saw Amelia Rosner's presentation at the Sixth Great Obituary Writers' International Conference. Cool and knowledgeable as she appeared making her speech, she might have been taken for an academic in

other circumstances; but at the end of the conference, after
one of her compulsive visits to the hotel computer, she had
run back, flushed with a kid's enthusiasm, shouting the news
of Reagan's death. Her standing in the frontier of obituary
studies is secure: *Amelia Rosner, who broke the news of Reagan's
death to a convention of professional obituary writers . . .*

Rosner's posts appeared in alt.obituaries every day, with six
or eight obituaries cut-and-pasted from papers around the
world, usually without comment, though sometimes she
would flag something special: "excellent & quite sad," she
wrote after one obit subject's name. She favored the British
writers, who were both looser and more intellectual than the
Americans.

Others on the newsgroup posted obits as well. One made
sure the AP wire summaries of deaths of the day got posted,
another kept up with dead sports figures; one followed military deaths, one kept up with dead lords and ladies; Bill
Schenley posted great obits from the archives; and so on. Between their posts and Rosner's, you could read your way
around the English-speaking world. Obituaries from
Malaysia, Australia, Pakistan, and Scotland were exotically
sprinkled in with London and a variety of North American
cities where the local papers ran feature-style obits, of either
prominent or ordinary people—New York, L.A., and San
Francisco; Washington, D.C., Toronto, Chicago, Boston, and
Houston; Minneapolis, Cleveland, Denver, Baltimore, and
Seattle; St. Petersburg, Austin, San Jose, and Sarasota. The
datelines read like an airlines departure board, which they
were, in a way.

In one brief period on AO, as the regulars called alt.obitu

aries, I read a streak of obits about people who had survived terrible chapters in history, beginning with one of the original Freedom Riders in the civil rights movement whose "life nearly ended 43 years earlier on a Greyhound bus in Anniston, Ala., when a mob firebombed the vehicle and held the door closed so the passengers—young people defying segregation laws as they rode through the region—would be unable to escape." This was followed by a woman who had been injured in the deadliest airplane crash ever, in the Canary Islands, then had lived another twenty-four years, and a man who had been marked for extinction sixty years ago for singing to Jewish inmates in a concentration camp cell block on the eve of the Day of Atonement.

The obits that families pay for, which resemble classified ads, are usually full of dreary repetition—*beloved . . . we'll never forget . . . the best mom in the world*—but feature-style obits are hotbeds of novelty. On alt.obituaries I read about people with wildly creative résumés: the former timber importer who became the Beatles' Mr. Fixit; the runway model who became a cop; the microbiologist who became an actress. Gathered in AO, as if filling the last bus out of here, were the odd ducks: the scholar who wrote *Reading the Bible in the Run-up to Death*; a Russian ethnographer and expert on skinheads; and Sir Dawnay Lemon, one of the last cricket players "to bowl under-arm seriously." (Bowling under-arm, you might be curious to know, is "now employed in serious games only as a cynical gesture against slow play.") I couldn't imagine more absurdly useless information, but I felt like Alice tumbling down the rabbit hole, one vivid character after another flashing past.

Besides the obits, plucked from websites around the world,

there were personal notes from people who had followed the subjects' careers, or worked behind the scenes on their last movie, or jammed with them in a roadside bar in Tarzana. When Danny Sugerman of the Doors died, an obit mentioned his wife, Fawn Hall. Was that *the* Fawn Hall, Ollie North's shredding secretary from the Iran-Contra scandal? Indeed it was. One of the posters dug a news story from "the bowels of Googlecache" that reported that Hall and Sugerman had tried to sell a book manuscript about their drug addiction called *My Heroin Honeymoon.* "Sugerman said in an online chat on 29 January 1996 that 'We are very much in love, living together with our dog, Bunky, in the Hollyweird Hills. In retrospect, we shouldn't have gone to the Golden Triangle for our honeymoon, however.'" You couldn't read that in the *New York Times* or even the *Daily Telegraph.*

The chatter around and between the obits on AO was like the chatter on most unmoderated, uncensored groups—obscenely freewheeling. Some of the liveliest discussions were off topic, and recklessly nasty. Never mind the political invective, which was everywhere in the contentious election year of 2004. Heated arguments erupted spontaneously over such things as whether or not deep vein thrombosis could be considered the third-largest killer disease in the U.S. Was DVT the same as stroke, or a subset of stroke?

Revelations made in candid moments came back to haunt the people who posted, and the tone was often personal and malicious. If people don't want their comments to be archived, they have to indicate this when they post, or what they write will live forever. One poster, in a moment he must surely regret, wrote, "I have never found diapers big enough for me." I know this only because it was cut-and-pasted and thrown back at him over and over again. *Diaper boy*, they called him, but he was hard to pity, because he flung more than his share of puerile invective. Anyway, pity in this arena is reserved for the dead, and sometimes not even for them.

Everyone seems to have at least one fellow poster whom he hates, passionately, and several posters claim from time to time to be stalked by their nemeses—so much for that stalk-free environment I was originally looking for. I felt a Darwinian indifference begin to seep in.

And, truthfully, I was entertained by the rants and scalding insults. As Schenley, one of my guides, pointed out, mean and nasty comments were not just tolerated but encouraged—as long as you were amusing. "Entertain us," he said, "or shut the fuck up." One scurrilous exchange came after someone posted a story about his father, a minister, who had swiped a plate of cold cuts after a wake. What a hypocrite, somebody pointed out. "Fuck off, you whorehouse miscarriage . . . ," came the vicious response, and he followed it up with, "go look in your own mirror, you monkey-spanked freak."

I emailed the dialogue to a friend who had just joined a newsgroup on a different topic. That kind of rant was called a flame-out, he said, enthusiastically. "It's so much fun!" (I noticed later, when I Googled "whorehouse miscarriage," that my friend had hurled the phrase himself on his own newsgroup, one devoted to the sophisticated discussion of wine.)

I read the postings on alt.obituaries day after day from behind a curtain of anonymity, and kept my comments to myself. I was intimidated even when no insults were flying. Geeks abound on AO. Facts, like opinions, are tossed about like glitter in a disco. If a singer died, a pack of posters would gather, disputing the details of her obit, recalling the time they heard an obscure version of her sixth-most-famous song, discoursing at length about the record labels and session musicians involved. They knew all the backup singers! It was a tower of factual Babel to someone like me, in the early stages of the Deep Forgetfulness.

"Was this the John Victor Monckton born October 1955 who was a great-grandson of the second son of the next brother of the ancestor of the Viscounts Monckton of Brenchley?" read one actual query, and out of the whole universe of obsessions came a breathless reply: "Yes. . . . The deceased is a fourth cousin of the second Viscount Monckton of Brenchley." I pictured the two royalists sitting in separate dusty libraries, tapping out their comments with little snorts of pleasure, their questions and answers sliding into AO history, below "monkey-spanked freak" and above the heated discussion of Veronica Lake's forgotten charms (and ashes).

The baseball addicts updating the list of the Oldest Living Major League Ballplayers spoke an equally impenetrable patois, based partially on numbers. They'd be huddling, not in the library but in some remote bleachers over their scorecards. "My count currently shows Bongiovanni at #21," one declared after the death of a major-leaguer forced a revision of their list of old survivors. "I have Harry Boyles (LOL) born 1/29/1911, at #21," the second little old man countered. Ah-ha! After many blustery posts, it turned out one of them was using the print almanacs and encyclopedias, and the other was using online references.

Yes, I know, they sound like a couple of box-score freaks,

splitting hairs in a Rotisserie League of the Doomed. But they weren't making a list of the best hitters or the best base runners—they were ranking the oldest ballplayers, the ones most likely to die. The obit culture has mated with the Internet culture, and this is the result: people who spend their time making catalogues of the near-dead, monitoring the critical-care wards of the news, watching spores of pneumonia drift and settle, scratching out the names of the losers.

The man who tumbled off the roof of a theater where he was working into a vat of boiling tar died a terrible death, which was acknowledged soberly on AO. "I would rather be hit by a train," one poster shuddered. "That had to hurt." Another commented, sympathetically, ". . . most likely he was 'dead' long before he hit the ground . . . if he knew he was going into the boiling vat of tar, he stroked out long before he was par-boiled. The mind is wonderful in that respect."

I don't know; do you find this comforting? I find this terrifically comforting. This is why I'm spending four hours a day on AO, I tell myself.

"I work on asphalt tankers," the next comment began, then the poster told the tale of a friend who died when he fell into an empty tank. It was sobering—perhaps too sobering. Someone observed that the accident caused the theater where the men were working to postpone its production of *Tale of the Allergist's Wife*. "Could have been worse," he speculated. "Could have been *Hunchback of Notre Dame*." And with this sly nudge, the gates flew open, and out poured the tarred possibilities: *Cat on a Hot Tin Roof. Song of the South. Horsefeathers.*

Reading obits and commentary on obits day after day, I came to consider this exchange fairly restrained. After all, there were several sober comments before the monkeys began flinging poo. The Sara Lee executive who was murdered and left in a deep freeze got no respect before the group pounced. The *nobody-does-it-like-Sara-Lee* and *just-warm-him-up-in-the-microwave* jokes began immediately. The sympathy came later. "I know it's part and parcel of how this NG [newsgroup] operates, but after a bit, I start feeling glad I'm not in this guy's family—not just for the loss, but for the inevitability of the jokes. Ultimately, it's somebody's father-brother-husband-son who's not so poppin' fresh anymore." The thoughtful author of this comment wondered about the day he dies, where the jokes will come from. Probably from his last name, he guessed, which was Beaver.

On AO, this was the equivalent of an engraved invitation. Just don't die in a bordello, suggested one of the buzzards swarming over his post.

I emailed Amelia Rosner questions from time to time about the shorthand and protocols of AO, and she graciously responded. She wrote that she had several correspondents who, like me, were lurking. "Have you posted yet? What are you waiting for?" she ended one email.

Lurking! I felt like a creep, suddenly. I felt, actually, as I had when I lived in New York City and subscribed to the *Geauga Times Leader*, the newspaper of a small town in Ohio where I'd once lived. For months I eavesdropped on my old friends and neighbors without participating in their lives. When the paper was delayed in the mail for a week, I called up the circulation department. *Oh, yes, the subscriber from New York City!* they said. Anonymity is relative. Rosner was flushing me out.

Soon after, someone complimented her online for all the great obits she'd been posting. "Thanks!" Rosner responded. "I imagine millions of lurkers reading them. (Otherwise, I wouldn't bother.)"

Lurkers—that word again. "This is as good a time as any to quit lurking and jump in," I began, and posted my first comment. Soon, and for a few brief glorious months, I was one of AO's ten most frequent posters.

I spent my workdays fishing for obits on the web, posting everything from a warm obit for a young transgender man in San Diego ("Whether as Angie or as Nathan, [the parents] said, he was still their adored middle child") to a rollicking British obit of the travel writer Pete McCarthy, who found humor in the transformation of Ireland from a backwater to a "rapacious tourist trap." (On a search for authentic Irishmen, McCarthy had been sent where the natives "ingeniously escaped the [tourists] by hiding in the last place anyone would think of looking for them"—Mac's, a fake Irish bar in a theme park.) I used a trick one of the posters shared with me: to type in "dies" or "died" in the Google search box and click on news. What I got, with the occasional great obit, was page after page of death stories, murders, accidents, seven Japanese who had met on an Internet newsgroup devoted to suicide and got together to fatally inhale carbon monoxide. You know—death. And obituaries, as anyone who reads or writes obituaries will tell you, are really not about death. They're occasioned by death, and they almost always wrap up with a list of those separated from the beloved by death, but they are full of life. The good ones are as intoxicating as a lung full of snowy air, as clarifying as the glass the ophthalmologist drops before your eyes, that brings the world into sudden sharp focus. The great obits aren't the products of jackknifed tractor-trailers and hurricanes—the obits are released by such disasters.

Googling death itself was a little depressing, frankly. I returned to my method, a hunt-and-peck equivalent in which I went, site by site, to newspapers that published feature-style obits, and read their current issues. I kept index cards filled with website addresses for newspapers from the *San Jose Mercury News* to the *Independent* in London, and visited them nearly every day. My goal was to find at least three obits a day that had a good story, vivid details, and were not just reported, but written with some flair or touch of artistry. I looked particularly for wild stories buried in otherwise ordinary lives.

Michael Taylor's obit of a Russian émigré in the *San Francisco Chronicle* contained everything I was looking for. Alexander Presniakov, eighty-nine, had been an engineer for Bechtel. He had also, as a small child, survived the Russian Revolution. His father, an officer in an elite royal cavalry unit, had fled with his family from the Bolsheviks

. . . through the winter night on a sleigh, racing across the ice and snow for the port of Murmansk, one of the outposts still held by the [White Russians]. At Murmansk, the captain of the last icebreaker held by the loyalists had drawn up the gangplank and was about to put to sea, as terrified refugees, stranded on the pier, pleaded with him to let them aboard.

"My grandfather took my dad and threw him over the water
and into the hands of a man who was standing on the deck of
the icebreaker. . . . The man went to one of the ship's officers
and said, 'Look, the child is here, the family is down there. We
have to let them on.' The officer finally dropped the gangplank
and let them on."

Where else would a story like this surface in our world? It
wouldn't be on the local news because there's no video
footage. It happened long ago, to someone who died, so we
won't be reading it on the front page, or the editorial page, or
in the lifestyle pages, where cookie recipes meet movie reviews. Only the obituaries keep such personal history alive.

And what a bonus to find this history thrown in with your
fifty cents' worth of box scores and bad news and celebrity
sightings. I even loved this particular obituary's anticlimax:
sometime after his historic escape from Russia, Presniakov
had worked in upstate New York in a plastics factory. The
plastics factory gave the whole obit perspective, and made the
dramatic rescue from the Bolsheviks real—that mix of the exotic and the homely, the crisp, fresh snow the sleigh skimmed,
mingling with the stink of melting plastic.

I knew immediately when I had found a winning obit. The
Associated Press story about Granny Plant, who had died at
111, the oldest woman in Florida, hadn't been fancy, but it
had the million-dollar detail: she'd come from Alabama with
her ten brothers and sisters by covered wagon. Dr. Lyle
French, who died at eighty-nine, had made his reputation
fifty years earlier, successfully removing half of several cere-
bral palsy patients' brains; believe it or not, removing half a

brain improved the patients' ability to walk, or so reported Ben Cohen of the *Star Tribune*. I posted an obit of a former semipro baseball player, Pete Grijalva, who had spent the last of his ninety-seven years on a barstool at "The Ould Sod" in San Diego, reliving the day he played an exhibition game with Babe Ruth and Lou Gehrig.

I think of it as harvesting. I'm sitting at my computer, in an office over my garage, now overrun with obit collections and stacks of torn newspapers and molding teacups. I can't start my day without cycling through my online newspapers, gathering the fresh obits, reading all of the posts. I want to tear my hair out when my computer is slow or my Internet connection falters. I bookmark the sites I visit most frequently, the big megillahs, *The Times* of London, New York, and L.A., and the *Washington Post*, the other London papers and the papers from Toronto, the *Star* and the *Globe and Mail*; the dozen or fifteen medium-sized papers; and my little favorites, the *Orange County Register*, where Robin Hinch writes about liquor-store owners and other ordinary people, and the *Point Reyes Light*, where Larken Bradley writes her pungent farewells to the flotsam and jetsam of northern California, and the *Las Vegas Sun*, where Ed Koch writes delicately worded farewells to old strippers and gamblers. (I posted one of a "casino greeter" and Rosner shredded both the obit and me; a simple Google, she complained, would have revealed the guy's gangster connections, omitted from the obit.) Access to all of these is free, though most papers require a simple registration and a log-in with a password. Some days, I find three in a row, bing bing bing, one great obit after another. Other days, I am combing the country of the Web for hours, looking under bushes, in caves, behind hillocks. Give me a beautiful obit. Just one beautiful obit.

I use only material that can be found free online. Part of the fun of posting is sliding the cursor down to capture a found text, shading it blue, then lifting it and arranging it in the blank, pulsing box. What shows up in AO is only the text, though some posters also include the Web address, in case readers want to go back to the source to read it in classier type, or check out the photos. One frequent poster on AO refuses to cut and paste the text. Instead he scrupulously posts the Web addresses, along with the official "U.S. and friendly nation laws prohibit fully reproducing copyrighted material. In abidance with our laws this report cannot be provided in its entirety. However, you can read it in full today at the supplied URL. . . ." My heart sinks when I click into AO and onto one of his topics, only to reach another address instead of a block of text, another set of gates that I have to register to pass through—not to mention that good-scout message. I've become so fast at whipping around the corners of my shortcuts, and I'm having so much fun, that his sober reminders seem like a traffic-court lecture.

As for copyrights, my magpie thievery is mild stuff. No money would change hands, anyway, and on AO, at least a few dozen people will see the work, attributed to a writer and a newspaper, without seeing the annoying flashing ad on the paper's website. On the other hand, some of those people will follow links or track down a source, and be introduced to the flashing ad. At any rate, in the case of obits, the layers of ownership are complicated. Whose material is it, anyway? The real person who created his life and then died, or the writer of that life—or the publication that pays the health benefits of the writer of that life? An obit is a palimpsest, and by the time it gets to AO, its form is as important as its content. An obit without an author or publication credit is missing vital information, like a CD without liner notes, or an individual player's box score detached from the rest of his team's stats. Is this a *Los Angeles Times*'s obit by the wonderful Elaine Woo, or found gold from the *St. Louis Herald-Tribune*? I put on a new title line. "Rev. Mark Poole, Nader Supporter." "Louetta Kambic, Weaver & Clogger." "Frances House, the James Bond of Priests." Like Rosner and some of the other AO posters, I'm a disseminator, spreading the nectar.

What is it we do when we post? We create cut-and-paste art. Like collagists, like junior Robert Rauschenbergs or Picasso manqués, we use scraps of found newsprint and our own comments to frame and present our discoveries. The casual, open, interactive gallery of Usenet not only displays this art but catalogues it as well. Do others want to see what we've cut and shaped? They can go to the white box in the upper right hand corner of AO's current page, the one marked "Search this group," and type in "Alexander Presniakov," or "The Ould Sod." In a flash, a complicated work of art appears, one that starts as the personal art of someone's life and history, continues as an artful piece of writing about that life, and lives on, selected and framed for our appreciation by a poster.

This artistic expression might actually be of use to someone who wants to compare the first drafts of the history of Katharine Hepburn or Ronald Reagan, but for the most part it is simply art, multilayered play that nourishes the soul. The best, I find, are made out of humble and unlikely material, like the obit of Suzanne Kaaren, ninety-two, an actress who had appeared in several Three Stooges shorts. In her obit for the *New York Sun*, Stephen Miller had crammed her identification with the fascinating particulars of her life: "an original Rockette, a champion high-jumper, a patent holder for a pop-top can, and in the 1990s [she] waged a successful legal battle against Donald Trump when the developer tried to evict her from her sprawling, rent-controlled Central Park South apartment." Miller, an amused sponge of pop culture, had sprinkled her obit with deadpan sentences like, "The Stooges seemed to value her opinion, and regularly tried out new material on her." The casual reader might have missed it, but Rosner was the poster. Over the cut-and-paste, she commented: "I would love any obit with the sentence, 'The Stooges seemed to value her opinion.'"

There is something about reading these obits, framed by such comments and trailed by wisecracks, that adds layers of appreciation. It's reading an obit with a gang of equally obsessed obit lovers; it's reading an obit with the marginalia intact. The gossip doesn't have to be phrased for a newspaper's legal department. The jokes are up-front. The corrections and editorial comments are part of the reading experience, and so are the personal notes from those who know something more about the subject. In short, the posted obits bring you into the bustling business of obituaries.

When James Fergusson, the *Independent*'s obit editor, complained to me about the cultishness that had developed around obituary writing and reading, he added, pointedly, "like some of these strange discussion groups . . ." I fought to keep my eyebrows from jumping while he trashed the obits scene on the Web. In fact, he didn't mean *these discussion groups*—he meant alt.obituaries, "the one that you occasionally appear in a lot—Google."

What a curious experience it is, being called cultish and strange by an obit editor. But what could Fergusson have against alt.obits? "Very often our obituaries are posted in that group and indeed elsewhere, without any acknowledgement, which is very irritating," he explained. Amelia Rosner posts the *Independent* almost every day, and every blue moon neglects the citation. It's easy to do if the newspaper's name isn't repeated after the byline and you have to type it in yourself. Rosner usually does this in the header with a compliment, or over the text—"fantastic *Independent* obit, really terrific, GREAT." She once posted an *Independent* obit that she said was the "kind of obituary you get when you take an extra day or two and you really want to do a superior job."

The *Indy* editor admitted he had seen these compliments. "She writes very cheery little things—I see that she occasionally puts comments." And he appreciated their impact. "Even with a comparatively small circulation, *The Independent* punches above its weight, as the English expression has it . . . just by being there on the net. Even though so many newspapers charge for archived material, if people are alert, as so many of these discussion groups are, then they can steal stuff in time, and it's then preserved."

AO's essential nature as a hijacking entity is bound to bother an obit editor. The posters don't solicit this editor's permission before disseminating his work; Rosner and all the alt.obits fans didn't ask his cooperation to install him as king of the obits. But that help-yourself grab bag works both ways.

"I look at it every day," Fergusson said, "because it's often an efficient way of discovering the death of somebody you didn't know had died. I can't say I have entered into all the controversies. A lot of those people, I just want to know what they do the rest of the day. A lot of these people make me feel absolutely desperate."

Of course they drive him mad; they drive all of us mad. But, what's this? *He's a lurker?!*

The man can say and do whatever he wants as far as I'm concerned, because he is a brilliant editor. A reserved, bookish man, he is also a pragmatist at a struggling newspaper with no library. Alt.obits is useful. He doesn't have to embrace it, or resolve his mixed feelings about seeing his work on its pages, or even tip anyone off to his presence. He can just drop in and take what he needs.

So, let's be perfectly clear about this. Alt.obits is populated by thieves and ghouls. It's also a boon to professional journalists. It is Grand Central, the next stage in the obituary revolution, the messy frontier in the great obit expansion. It's a place where readers can become critics and editors can monitor rumors of death. The writer from the western paper posts an obit her editor wouldn't let her run. The webmaster of the site www.deadpeople.info verifies a death. As for the question of whose obits are better, it's easy to see when comparisons are but a keystroke away. The result is a sharing of resources, a leaping of influences. The beating heart of a living art form is being tracked on the Web.

11

The Obit Writer's Obit

W. C. Fields died on a Christmas day; so did Charlie Chaplin and Dean Martin. Christmas 1989 was memorable: the Ceauşescus were shot, and Yankees manager Billy Martin's drunk friend killed him, driving his truck off the road. Christmas 2004 was relatively peaceful, until the disaster.

Someone on alt.obits posted an early alarm at 3:15 A.M. on the twenty-sixth, and news of the devastation in the Indian Ocean began to filter out: an 8.9-magnitude earthquake followed by a tsunami of unthinkable proportions. Early estimates of the dead began doubling, then kept doubling until more than 150,000 people had been crushed, drowned, swept away. Articles, eyewitness accounts, body counts, and obits of the individuals lost began filling the columns of alt.obits.

Against this backdrop, even deaths unrelated to the catastrophe seemed charged and poignant. Wildlife artist Simon Combes was killed in Africa, and his obit in the London *Independent* included links to some of his paintings, so detailed I thought they were photos. One pictured a lioness leaning against her mate, and beside it Combes's recollection of having seen the lions couple moments before. Reading it, I had that elusive feeling of being alive in all my senses. His obituary was a reminder of the possibilities in life: a career as a voyeur of lions, and a death worthy of an adventurer—gored in the bush by a wild buffalo.

The contrast of this obit appearing amid the rubble of the great disaster struck me. The tsunami, as it turned out, would strew wreckage in newspapers and obituaries for months. Its dimensions would be measured in height of waves and number of deaths, but it was only in the particulars, the mother who clung to one child while another washed away, that I could grasp any of it.

It was school vacation; I had a houseful of children and guests; but I felt the need to add to the chorus, to find one obituary—any obituary—that might chase away the despair of mass tragedy with a tribute to a particular, singular life. I started as I often do, at www.philly.com, to check out Yvonne Latty at the *Philadelphia Daily News* and Gayle Ronan Sims at the *Philadelphia Inquirer*. Neither had obits that day, but I recognized a name in the bold headlines: Leon Earlen Taylor III, the obit writer who had taken over for Jim Nicholson. I had interviewed him twice over the phone, and the second time, a month earlier, I learned he was ill; but talking in his deep, resonant voice about the satisfaction he got walking people through their grief, he hadn't sounded sick. Taylor had covered a lot of ugly deaths as a reporter, including the MOVE tragedy, when Philadelphia police had firebombed the house of a black revolutionary group, killing eleven and burning a city block. Writing obits of ordinary people who lived lives of quiet pleasures had been a refuge for him, and the joy he took in it came through clearly.

I wish I'd known him is the response every good obit writer tries to elicit; it's a universal marker of obit success. But what if you already know him? The news overwhelms everything; obits lose their cool, almost campy appeal. Who cares about resonant details? It's hard enough to read through tears.

Is an obit writer's obituary sadder than anyone else's? Poems about poets, plays about playwrights, films about film-makers—all, at their best, achieve a kind of harmonic resonance, layers of irony intensified by the concentration of subject and form. An obit writer's obit has the same effect. Why? It's just an obit—an announcement that, no matter what else it accomplished on this earth, a body had been separated from its spirit. And yet, this obit has a disorienting spin. The byline of the writer isn't anchored at the top, in control, but trapped in the body of the story. *There it is*, you say to yourself, putting down the paper or turning away from the screen—*death, the great equalizer.* Even the gatekeepers have to go through the gate.

Any death that happens to someone you know is going to resonate more than the death of someone you don't. There's death, and then there's death. In the novel *Who's Who in Hell* by Robert Chalmers, a former *Daily Telegraph* obit writer tells the story of an obit writer for a newspaper strikingly similar to the *Daily Telegraph*. Every death is the occasion of a piece of writing, an opportunity to render colorful details and entertaining quotes, until someone close to him dies. He and his editor huddle in tears. *Oh, dear*, jeers a fellow journalist, *does the obit page have a death?* The obit editor is furious. *You don't understand*, he says. *This is a* real *death!*

Taylor's obit in the *Inquirer* was as straight as they come. "The *Inquirer* is the 'paper of record,' and their staff kind of carries that attitude," he had told me.

> *Leon Earlen Taylor III, 52, a former reporter at the Philadelphia Daily News who as an obituary writer helped families capture memories of their departed relatives, died Sunday of lung cancer at his home in Middleburg, Fla.*
>
> *Mr. Taylor, who retired three years ago, began his 25-year career at the Daily News in 1976 as a copy boy. He was later a reporter with beats including general assignment, welfare, neighborhoods, and obituaries.*
>
> *During his six years writing obituaries for the Daily News, Mr. Taylor wrote moving accounts about everyday people, such as the "librarian with a smile that lit up a room," the "probation officer who was a class A dad," and "a cop's cop."*

On the other hand, Taylor also told me, the *Daily News* was "the dirty news. We're a lot more down to earth." Although the *Inquirer* and the *Daily News* cover the same city, share the same office building, and are both Knight-Ridder properties, nobody confuses them. You can boil their differences down to one detail: the *Daily News* obits use nicknames. Nicholson began the custom, arguing that if you wrote an obit about Thomas Robinson, everybody who'd known "Moose Neck" would miss it; and Taylor enthusiastically continued it. "One of our standard questions was, 'Did he have a nickname?'" He always asked. So if you didn't know otherwise, the *Daily News* would tell you.

LEON "THE FLY" Taylor, a retired Daily News reporter whose stories carried credibility and whose life centered on compassion, died Sunday of cancer. . . .

In the 1970s, when the "Superfly" movies hit, Taylor got his nickname. He fancied the leisure suits and bell-bottoms and broad-brimmed hats. In later years, all that in-the-groove vogue gave way to the man who was growing inside. The 6-3, 200-pounder with the close-cropped gray hair and graying goatee wore perfectly tailored double-breasted, three-piece suits and color-coordinated tie and shirt, usually with cuff links.

Taylor striding into the newsroom always looked a bit incongruous at first. He had a cinematic presence sort of like a cross between Morgan Freeman and Richard Pryor. Actually, the man could have been anything . . .

There had been no byline on the *Daily News* story online, but I should have known. The newspaper, which ran the obit with photos and a reprint of his Viagra story over three pages, ran its attribution in a box, which soon appeared on the website.

Jim Nicholson was an award-winning obituary writer for the Daily News for 19 years. His style has been featured in journalism texts and widely copied. He was Leon Taylor's obituary-writing mentor, and has come out of retirement to write Taylor's obituary at Taylor's request. . . .

"I got a call yesterday from a clerk there who told me of his death. I thought it was just a courtesy call," Nicholson's email said. "But, then he transferred me to Yvonne Latty. She said that just before [Taylor] died, he asked his wife and his best friend Foster to make sure I wrote his obit. I got a lump in my throat over that, pard. So all day yesterday I was fed interviews and data from Latty and Barb Laker (a close friend who sat next to him). I wrote from 2 pm to about 5 pm and made deadline."

The Fly had been a self-driven copyboy, determined to make it as a bylined writer. He got a degree in journalism from Temple University and achieved his goal: he got to cover "fatal fires, car crashes, attempted ax murders, kidnappings, homicides—some real freaky ones, including the guy who snatched women off the street, then killed and cooked one in pots on the stove and fed her to the other women . . . a lot of juicy things like that. I did a few undercover first-person pieces, worked as a day laborer picking blueberries in New Jersey, lived on the street as a homeless person.

"After fifteen years, I was burnt out, turned off from writing, when I started hanging out with Jim and looking over his shoulder. It didn't take him long to convince me that obits were what I needed to stay sane.

"I've had years of tearing families down. I was the one who said, 'It was your son who set the fire that killed that five-year-old girl . . . your kid is the one who ran over that eighty-year-old couple last week.'" As an obit writer, Taylor called after the bad news had already been broken; his job was to help people remember the good times. The process was cathartic for the survivors—it breathed life back into the dead—and it helped heal the journalist, too. The man who'd seen too much gruesome fallout was gifted at handling ordinary tragedies.

The key, Taylor told me, was being *low*-key. "Everyone who comes in the house or calls on the phone is trying their best to out-mourn everyone else. You get a steady diet of that for three or four days, and I call up and talk regular, just like I'm talking to you, it's a breath of fresh air. I ask about his favorite breakfast food, what kind of disciplinarian was she, did she let you have that hamster or dog the first time you asked. I'd wind up asking the questions that people who live with you all your life never end up asking. And then I'd hear, 'First time I've been able to talk about this without crying.'"

Instead of making people weep, Taylor made weeping people smile. "I'd walk away from the day feeling positive about what I'd done. I'd be laughing and joking, falling out of my chair. 'What were you doing?' somebody would ask. I was doing an obit interview."

From obits written by Leon Taylor:

She started out with just a hot comb and a dream in a North Philly project.

If you wanted to keep your secrets secret, the last person you would want to share them with was "Sonny" Moody. He meant well. It's just that things sometimes slipped out. Like the time his brother gave his new girlfriend a ring that had been returned to him by his old girlfriend.

You don't normally think of Irish dancing as a contact sport.

He sounded like Nicholson around the edges—"I learned at his knee." Both writers could turn sappy; it was the risk you ran with egalitarian obits—or maybe it was the point. Even more than Nicholson, Taylor felt it was his job to take care of the families. His embrace was literal. When people stopped by to drop off a photo, "I always greeted them, I always gave the women a hug and a kiss on the cheek, a hug and a handshake to the men." He was a kind of minister, laying his hands on the survivors.

"Nobody wants the beat!" Taylor marveled, and yet to him, obituaries are the best job in newspapers. When he needed to help care for his father-in-law, who had Alzheimer's, the *Daily News* let him write from home, and he used that fact to persuade the young up-and-coming city reporter, Yvonne Latty, to step in as his replacement. "Yvonne had just had her first child, but she hadn't been around long enough to see the advantages in obits. She had her career path planned out, she wanted to go up that corporate ladder, obits maybe at the end. 'You will fall in love with this job,' I told her. 'You'll be able to stay home, you'll feel good about yourself every night.' I talked and talked and talked to her, and finally she took it. I stopped down there a couple of times, and she said, 'Man, the best move I ever made.'"

The obits desk at the *Daily News* is so successful, "every day I had to turn away one or two or three families because we didn't have space," Taylor said. When he retired and moved to Florida, near Jacksonville, he decided to try to tap that demand by freelancing, figuring that "instead of flowers, people can commission an obit that would be part of the family's legacy and heritage for decades to come." The problem was "the kind of obits I write are not in their culture. People think I'm talking about death notices." He targeted twenty or twenty-five newspapers that ran what he called "living obits" like the ones he'd written for the *Daily News*, papers like the *Detroit Free Press*, the *Baltimore Sun*, the *Star Tribune* in Minneapolis, the *San Diego Union-Tribune*, and he sent potential customers to the newspapers' websites to see what it was he was selling. He died before the egalitarian obit could take root in Jacksonville.

The altar was a simple dark red velvet curtain with a cross; the walls were cinderblock covered with strips of wood paneling; a red carpet ran through the center of the Redeemer Moravian Church in Southwest Philadelphia; otherwise, the floors were green linoleum, spotless and glazed. I had made my way there by a combination of trains and buses, unsure of where I was going or how long it would take, so I arrived early, and had lots of time to study the program, a picture of Leon Taylor on the cover in a sharp suit, holding a cigarette to his mouth. Watching the pews fill with black and white, male and female, old and young, I thought of the line from Nicholson's obit of him: "In the newsroom, where age and sex and status create the inevitable office cliques, Taylor moved through the social strata like it was vapor." I wished Nicholson could be there, but by then he was taking care of his fading wife and his mother, who was in her eighties and had heart problems. I picked up a Bible from a stack at the end of the pew, *The Good News Bible*, the plain-talk Bible. I turned to Ecclesiastes, which Nicholson had extolled. "Life is useless, all useless. You spend your life working, laboring, and what do you have to show for it? . . . It is like chasing the wind. . . . Enjoy every useless day of it because that is all you will ever get for your trouble. Work hard at whatever you do, because there will be no action, no thought, no knowledge, no wisdom in the world of the dead—and that is where you are going. . . ." Bleak and matter-of-fact—deal with it. Enjoy every useless day. I resolved to study the book when I got home.

Three white women settled in behind me and started talking about working with the Fly. Someone remembered his early take on the news office: "'Everybody talks about lunch for three hours, and where they're going to eat, then they go off to the restaurant of the day,' he said. 'I don't fit in! I bring my lunch.'" They remembered when he worked the night shift at police headquarters. According to these women, he had made life on that grim beat fun. Was he good? No question about it; there are five people now doing what he used to do. Their reminiscences were a kind of street-corner obit, like the ones Nicholson and Taylor modeled theirs on.

Taylor's wife and her sister settled in a front pew. His only child, a daughter from his first marriage, sat behind them with her cousins. Eight robed men and women, the Redeemer Moravian Church Choir, filled the space with sacred and soulful music, accompanied by an organ, the pride of the church. One soloist, then another, emerged from the congregation to sing *a cappella*, and then, after a dramatic start, the accompanist chimed in on the upright piano tucked behind the organ.

But all that could have been for anyone. The Taylor-specific portion of the program was a unique mix of off-color anecdotes, humor, and affection for a man at ease with men and women of all colors, and at ease with himself. The time Taylor took Viagra to write a feature about the new drug, one coworker recalled, he'd gotten a call almost immediately to write a last-minute obituary, which he polished off, in spite of the distraction. Someone else remembered the night the Fly went to cover a stakeout of a bordello. With the local TV cameras rolling, he'd knocked on the door—they thought he was one of the customers they'd come to bust, not another journalist, and he played them along. That was the Taylor who came to life in the church: bold, funny, and cool.

The obituaries were mentioned; in fact, there was a palpable tension over the obituaries. Taylor's ex-wife made an unscheduled trip to the lectern to tell us all that the *Inquirer* had done an especially fine job with his obituary, which had been a particular comfort to their daughter. The pews full of Taylor's *Daily News* coworkers shifted uncomfortably. The minister, a peacemaker, made a point of saying before his eulogy that he had not known Taylor personally, but felt he did because of the wonderful obituary by Mr. Jim Nicholson in the *Daily News*. Like almost all preachers, priests, and rabbis, he made liberal use of the obit in his eulogy.

The service ended, and we filed out while the choir moved us along, and the line of survivors captured what they could from us about Taylor before we melted away. The warm news on this cold night was that Taylor's daughter had been accepted at the University of Pennsylvania, the Ivy League school that looms like a fortress in the gritty neighborhood of West Philly; it was a bright affirmation of his success.

I had a bus ride and three train rides ahead of me, and it was nine at night. I asked several women in the ladies' room if any of them was driving downtown. The young woman who came to my rescue turned out to be Yvonne Latty, Taylor's replacement at the *Daily News*. She zipped over the expressways, and what had seemed a farflung neighborhood, I saw, wasn't far away at all. What had been in the *Daily News* that had so offended Taylor's daughter? I wondered. Nothing, *nothing*, she said. The *Daily News* hadn't mentioned the ex-wife; it's their policy to take the survivors list from the closest survivor, and the current wife hadn't mentioned her. The paper of record, on the other hand, had dutifully mentioned Taylor's ex. "That obit in the *Inquirer* was flat, just flat," Latty said. She found the *Inquirer* obits lifeless, as a rule. They were really no competition.

"Fly talked me into doing this," she acknowledged. Three months later, so inspired by the interviews she did on the obits desk, she began writing her book, *We Were There*, about African-American vets in a number of wars. It was an idea anyone on the obits desk might have thought up—someone should capture these stories *before* the people die—but Latty, dynamic and not yet in her thirties, had made it happen.

"I had been an urban reporter for years. I used to write murder stories, kids getting murdered, and I'd have to talk to the parents. I thought that was real. After writing obits about foster mothers, machinists, all these inspiring regular people, I realized murder isn't real life. The people in these obits are real life."

"If you want to talk to all the obituary writers in Israel," the man is saying, "you're talking to them." He holds up his hands with a shrug, as if to say, I know it's absurd. He is the only one sitting on the dais.

How is this possible in any country, but especially a country like Israel, crammed as it is with history and violence and eventful deaths? It's a missed opportunity that shocks the room full of obituarists, gathered in Bath, England, for the Seventh Great Obituary Writers' International Conference. They laugh in disbelief.

Uri Dromi writes two relatively short obituaries every week in *Ha'aretz*, the literate and liberal national paper of Israel. The newspaper is published in Hebrew; every few weeks, an English translation of the obits finds its way to the Web. Dromi does other things besides writes obituaries. He is in charge of outreach for the Israel Democracy Institute, and writes about Mideast politics for the *Miami Herald*; he's been a spokesman for Yitzchak Rabin and a colonel in the Israeli air force, where he served through four wars; but writing these short life summaries might be his best legacy. He sends off people like the woman pulled from the concentration camp oven's door who was then shelled mistakenly by her British rescuers but survived to start dental clinics for children. He admits, "My tendency is to write favorably." He says he is overwhelmed with emails from grateful readers. "Before, there were wonderful people, and they would just go away, and that's it."

[Akiva Nir] jumped from the window of the Gestapo office and escaped to Nolcovo, the village where his father was born. He was turned in to the Nazis on the eve of the liberation, but the village's mayor, Yan Dovovitz, swore with tears in his eyes that Nir was his grandson, and Nir was saved.

And Nir was saved again, one could argue, from oblivion, if not death, with this obit.

A year after Reagan's death electrified a wake of obituarists in New Mexico, the organization gathers in the ancient city of Bath in southwest England. This conference feels truly international: Besides the Israeli writer and a raft of Brits, there are Canadians, Australians, and even a Spanish scholar.

Spain, we learn, has three national newspapers publishing *obituarios*. Some of them are measured stories of a newsworthy person's life; some of them are opportunities for publicity-hounds to brag about their friendship with famous departed. The young doctoral candidate researching Spain's obituaries, Isabel Corona, explains the differences between the papers, and flashes pictures of their pages on the screen, throwing out observations and shrewd asides: the rest of us in the room are "writing to each other, no?" She has fallen into the conference like any obit-lover falls—disbelieving, and overjoyed to find others who share the passion.

The reason the Canadian national newspaper the *Globe and Mail* has such a lovely obituary page, we learn, is because Lord Ken Thomson, its former owner, who retains a stake, had a son who was a "gluestick/scrapbook collector" of obits, who hated seeing all that clutter. "It used to be a compromised page," says the *Globe and Mail*'s obits editor, Colin Haskin, "full of page turns, jumps, and an advertising banner." With one phone call, Lord Ken swept away the mess. Now these pages, which have to be seen in newsprint to be appreciated, are the prettiest in North America, with big vintage photos of subjects in their prime, a deep lead obit, a section called "Lives Lived" written by readers, and no clutter.

The Australian Dr. Nigel Starck, natty in his safari suit, sweeps through Bath enroute to Hong Kong, to talk about his research into the history of the obituary. He is drafted at the last minute to deliver the keynote speech when Hugh Massingberd, frail of health, begs off till the next day; and Starck not only cheerfully reaches into his briefcase to assemble the speech but invites me to sit with him while he makes notes. I've been trying to nail down the pieces of this story while events tumble past, but Starck has spent decades trying to decipher disintegrating fragments of newspapers and microfiche. "Would you like to see the very first obituary published in Australia?" he says, and pulls out a copy of an account of a funeral published in the *Sydney Gazzette* on the second of March, 1804, originally printed on homemade newsprint. Starck, something of an obsessive/compulsive, read every newspaper he could get his hands on in Australia, column by column, afraid that what he was looking for wouldn't be listed in any index. He knows for a fact that James Bloodworth was the first person on that continent whose death was noted in print. I copy out every word of the old story to give Starck a few minutes to compose the opening of his speech, and also to honor him. The indentations and typography and language are as odd as those in an Emily Dickinson poem.

Drum muffled & Fife
THE BIER
Two Sons, chief Mourners followed by an
Infant Daughter,
Fourteen Female Mourners,

Twenty-Four Male ditto,
A number of respectable Inhabitants in Rank
The Non-commissioned Officers of the New
South Wales Corps
And a crowd of spectators

Over three days, the obit writers and fans discuss their passion over eggs and blood pudding, or ale and nuts, in the hotel with a World War II bomb still buried in its bowels. The tension between those who write witty London-style obits and those who celebrate ordinary people continues, and finally erupts. One of the contributors to the British press disparages the way Americans waste space on "the foibles of nonentities and their pound cake recipes," and collectively, the people in the room suck in their breath. The obits editor of the *Atlanta Journal-Constitution*, Kay Powell, drawls, "Everybody in the South is eccentric to some degree, so what to you is foibles is normal to us." Then she tells a story about working on the obituary of the nonentity who owned a wine store in Atlanta. A little research turned up the fact that he had a back room behind his store where he would cook and serve food to go with the wines he was promoting. One day, Marvin Griffin, the former segregationist governor of Georgia, Ralph McGill, the liberal editor of the old *Atlanta Constitution*, and Dr. Martin Luther King, Jr., found themselves shopping in the store at the same time, and ended up in the back room together. A few bottles of Mr. Jim Sanders's wine smoothed the meal in this tantalizing footnote in the history of the South. The three men, all great storytellers, stood outside the wine store after it closed, laughing and swapping tales. King was killed soon after. The vast waterfall of history pours down, and a few obituarists fill teacups with the stories.

Jim Nicholson and I kept up a daily email correspondence for more than a year; he worked—he must have prayed—to get me to read Ecclesiastes. "Read the book, six and a half pages. Much wisdom. So fraught with reality, most preachers will not even include it in their repertoire of sermons. Too heavy for some folks. Starts out very dark, lifts, dips, ends well. Visualize King Solomon in an upper chamber, late at night, pacing before a small table where a scribe sits taking this down with only one candle casting a glow. The King in a burst of inspiration, unable to sleep, comes forth with the essence of his wisdom, his fears, his discoveries, his warnings, the promises, his depressions, his questions . . ." And later, Jim added, "The wisest man to ever live still cannot fathom some of life's mysteries, or injustices, even against the backdrop of God's love. His frustration is clear. It is as fresh and current as last night. But, the hope and promise are there for those who look for it."

I finally sat down and read it through, after dipping into it at Leon Taylor's service. I recognized the passage that begins "To every thing, there is a season, and a time to every purpose under the heaven," because the Byrds sang it. Ecclesiastes was a great book, I thought, for Nicholson—the writer who had quit writing, the man who met his fate indifferently, whether he was caring for an Alzheimer's patient, or writing an obit, or drawing up the profile of a terrorist. One mission was the same as another. It was useless, all vanity and vexation of spirit, just as the wise king said. "There is no man that hath power over the spirit to retain the spirit; neither hath he power in the day of death; and there is no discharge in that war." But it was also a great book for me. The second-to-last verse reads: ". . . be admonished: of making many books there is no end; and much study is a weariness of the flesh." Indeed!

"Some people read the obits to look for the secret to a successful or happy life. How did this person get through seventy or eighty years, what can I learn from it? It reminds us of those core values . . . religion, honor, loyalty," he had said. Nicholson's point was that we don't have to keep reinventing these things if we keep them alive, and we *can* keep them alive, in the obits.

The closest he and I ever came to an argument was talking about the man who couldn't pass the raised hood of a car without stopping to take a look. Nicholson thought people loved that detail because it reminded them of somebody they knew. I thought people loved that detail because it was original and unique. In my view, the whole point of Nicholson's profiles—and Leon Taylor's and Yvonne Latty's and the other obit writers who had helped democratize the obit—was to pin down the individuality of each one of those people leaving the earth in sheets and waves. I thought what was so special about their obits was their rescue of the individual from the fate of the masses. And Nicholson thought what he and the other obit writers were doing was reminding people of something they already knew—"there is no new thing under the sun."

I still think that the point of the obituary and the beauty of it, aside from its elegant structure and the wonderful writing it can inspire, lies in that heroic act. There goes *one*, the only one, the last of his kind, the end of a particular strand of DNA. Make whatever you want of the history and sociology that a run of obituaries shows us about Philadelphia in the second half of the twentieth century, or how that wave of obituaries changed this literary form, or how the explosion of candor in London has elevated the art of newspaper writing— what's most valuable about the obit, any good obit, is how it tries to nail down quickly what it is we're losing when a particular person dies, the foster father of two hundred children, or the "revolutionary whore" who wrote books and founded an international library of prostitution, or the woman who "loved to sew pillows and knit newborn baby caps"—all of them, any of them. The better the obit, the closer it approaches re-creation. It's an act of reverence, a contemplation of this life that sparked and died, but also an act of defiance, a fist waved at God or the stars. And what else, really, do we have besides the story?

Appendix

═══════

INTERNET TOUR GUIDE

The following is a list of websites I monitored in my research. It isn't exhaustive—the scene is dynamic, and the categories are often blurred—but it should provide a good base for further exploration.

THE LONDON STYLE

The *Daily Telegraph*, London
   www.telegraph.co.uk

The *Globe and Mail*, Toronto
   www.theglobeandmail.com/national

The *Guardian*, London
   www.guardian.co.uk

The *Independent*, London
   www.independent.co.uk

The *New York Sun*
   www.nysun.com
The Times, London
   www.timesonline.co.uk

NEWS OBITS
The *Dallas Morning News*
   www.dallasnews.com
The *Houston Chronicle*
   www.chron.com
The *Los Angeles Times*
   www.latimes.com
The *New York Times*
   www.nytimes.com
The *Philadelphia Inquirer*
   www.philly.com
The *San Francisco Chronicle*
   www.sfgate.com
The *Star Tribune*, Minneapolis-St. Paul
   www.startribune.com
The *Washington Post*
   www.washingtonpost.com (metro section)

"EGALITARIAN" OBITS
The *Atlanta Journal-Constitution*
   www.ajc.com
The *Baltimore Sun*
   www.baltimoresun.com
The *Boston Globe*
   www.boston.com

Chilkat Valley News
  www.chilkatvalleynews.com
The *Denver Post*
  www.denverpost.com
The *Las Vegas Sun*
  www.lasvegassun.com
    (search: Ed Koch)
The *Miami Herald*
  www.miami.com
Newsday, Long Island, N.Y.
  www.newsday.com
The *Orange County Register*
  www.ocregister.com
    (search: Robin Hinch)
The *Oregonian*
  www.oregonlive.com
    (search Amy Martinez Starke and Joan Harvey for their
    "Life Stories")
The *Philadelphia Daily News*
  www.philly.com
The *Pittsburgh Post-Gazette*
  www.post-gazette.com
The *Plain Dealer*, Cleveland ("A Life Story")
  www.cleveland.com
The *Point Reyes Light*
  www.ptreyeslight.com
The *Rocky Mountain News*
  www.rockymountainnews.com
    (search Jim Sheeler and/or Obituaries)

The *San Diego Union-Tribune*
  www.signonsandiego.com
The *San Jose Mercury News*
  www.mercurynews.com
The *Seattle Post-Intelligencer*
  www.seattlepi.com
Toronto Star ("Lives Lived")
  www.thestar.com
The *Washington Post* ("A Local Life")
  www.washingtonpost.com

TRIBUTES
The Atlantic Monthly
  www.theatlantic.com (search: Mark Steyn)
    (the website prints only his first paragraph or two,
    though, and you'll want to read the whole story)
The Economist
  www.economist.com

OTHER OBITUARY RESOURCES ON THE WEB
"Final Curtain." the obituary radio show
  www.kcrw.org/show/fc
Blog of Death
  www.blogofdeath.com
GoodBye!
  www.goodbyemag.com
Google: groups: alt.obituaries
  http://groups-beta.google.com/group/alt.obituaries
The International Association of Obituarists
  www.obitpage.com

Obituary Forum

  www.obituaryforum.blogspot.com

The Poynter Institute for Media Studies

  www.poynter.org

      (with links to articles about obituaries by Chip Scanlan

      and Trudi Hahn, Hugh McDiarmid Jr.'s profile of

      Robin Hinch, and other information on obits)

Wikipedia

  www.wikipedia.org

Notes

CHAPTER ONE: I WALK THE DEAD BEAT

The quote "Within the short period of a year she was a bride . . . a corpse!" is from the *Baltimore Sun*, August 3, 1855. "Death of Mrs. Captain Reed," as quoted by Janice Hume, *Obituaries in American Culture* (Jackson: U. Press of Mississippi, 2000).

CHAPTER THREE: NAME THAT BIT

The quotes from the obits of Billy Carter appeared in Robert D. Hershey, Jr.'s story in the *New York Times*, Sept. 26, 1988, and in Hugh Massingberd, *The Very Best of the Daily Telegraph Books of Obituaries* (London: Macmillan, 2001).

CHAPTER SIX: ORDINARY JOE

Quotes from the following obituaries by Jim Nicholson,

published in the *Philadelphia Daily News*, appeared in this chapter without context. Identification follows:

"They were married three months later . . ." from "Joseph Lees Sr., 59; a salesman all his life," June 18. 1991.

"It was one hot day . . ." from "Sarah Hanton, mother with 'magical' powers," August 14, 1991.

"In a lot of scenes . . ." from "Viola MacInnes, 66; headed foundation," April 16, 1993.

"Society today does not assign . . ." from Christopher J. Kelly; the joy of his family," April 14, 1993.

"He had the digestive juices of a shark," and "My dad grieved hard when women would die," from "Herbert Speach, 59; ex–post office worker," August 20, 1991.

"Charlie did it all with one eye," from "Charles M. Myers, 75, rough & tumble photog," August 22, 1991.

"I had unfortunately burned up my cat Smokey . . ." from "Louis 'Lou' Koreck, a many-faceted gem," September 29, 1997.

"It was the worst show I've ever seen in my life . . ." from "Stan Hurwitz Loved Shows, Friends," January 7, 1991.

"It would take him two minutes . . ." from "30-Year Cop James F. O'Neill," April 26, 1989.

"Will could be gruff at times . . ." from "'Will' Williams, Re-tired News City Editor, January 20, 1988.

Till Death Do Us Part: 101 Love Stories That Would Not Die, a collection of Jim Nicholson's obituaries, were gathered for publication by the *Daily News*, but never published. The *Daily News* charges for access to its archives; it costs $2.95 to read each obit. They can be read for free at the Philadelphia public libraries; two are reprinted in *Best Newspaper Writing 20th Anniversary Scrapbook*, and five are reprinted, with an interview, in Don Fry, ed., *Best Newspaper Writing 1987* (St. Petersburg, Fla., 1987).

CHAPTER TEN: GOOGLING DEATH

To read the posts in this chapter in context, go to Google, click on Groups. Type in alt.obituaries, then in the search box for this group, type in a name or key phrase. Or type any of the following into your browser for the context of some of the discussions I quote:

Bob Prince: http://groups-beta.google.com/group/alt.obitu aries/browse_frm/thread/8ec742cae15dc320/39870455b249c1 7d?q=bob+prince+bill+schenley&rnum=1#39870455b249c17d

Freedom rider obit of Ed Blankenheim, Oct. 3, 2004, by Wyatt Buchanan in the *San Francisco Chronicle*; Canary Island crash survivor obit of Bethene Miller Moore, Oct. 17, 2004, by Michael Taylor in the *San Francisco Chronicle*; concentration camp survivor obit of Joseph Cooper, Oct. 3, 2004, by David Grossman in the *Toronto Star*. Other specific obits can be found by entering the name or phrase in the box "search this group."

Danny Sugerman/Fawn Hall thread: http://groups-beta. google.com/group/alt.obituaries/browse_frm/thread/888ea4

a2ae763e09/5c06d25b4bc25513?q=danny+sugerman&rnum
=1#5c06d25b4bc25513

Diaper boy reference (one of many): http://groups-beta.
google.com/group/alt.obituaries/browse_frm/thread/71cfe2
df21e4a766/6bce41602fd64421?q=%22diapers+big+enough
%22&rnum=23#6bce41602fd64421

"Whorehouse miscarriage" thread begins at: http://groups-
beta.google.com/group/alt.obituaries/browse_frm/thread/27
d0433e24780e28/1fd1a38ac4682f1e?q=KY+wizard+father+
minister&rnum=3#1fd1a38ac4682f1e

John Victor Monckton: http://groups-beta.google.com/
group/alt.obituaries/browse_frm/thread/12d5c9917489a73d/
6289351080fc9ff9?q=John+Victor+Monckton&rnum=2#628
9351080fc9ff9

Oldest living Major League ballplayers discussion (one of
many): http://groups.google.com/group/alt.obituaries/
browse_frm/thread/b8b9bff9cb17814b/8ab83cd7d5a9fbed?
q=harry+boyles&rnum=2#8ab83cd7d5a9fbed

Boiling vat of tar thread: http://groups-beta.google.com/
group/alt.obituaries/browse_frm/thread/708452d0b573aaca/
12cae0a77dc56567?q=boiling+vat+of+tar&rnum=1#12cae0a7
7dc56567

Sara Lee executive found frozen discussion: http://groups-
beta.google.com/group/alt.obituaries/browse_frm/thread/
51243e2136abd4df/b3aa6eac0d136341?q=missing+sara+lee+
executive&rnum=1#b3aa6eac0d136341

Suzanne Kaaren and the Three Stooges: http://groups.
google.com/group/alt.obituaries/browse_frm/thread/501c4fc
a6135a48b/27033049c4890055?q=suzanne+kaaren&rnum=
1#27033049c4890055

CHAPTER ELEVEN: THE OBIT WRITER'S OBIT

"She started out with just a hot comb . . ." from Mildred
Garrett, 62, salon was her dream," by Leon Taylor in the
Philadelphia Daily News, September 15, 1999.

"If you wanted to keep your secrets secret . . ." from "Sonny
Moody, sports, sci-fi fan," by Leon Taylor, the *Philadelphia
Daily News*, May 25, 2001.

"You don't normally think of Irish dancing . . ." from "Kevin
Donnelly, earned Irish man of year award," by Leon Taylor,
the *Philadelphia Daily News*, August 2, 1999.

Bibliography

Ball, John C., and Jill Jonnes. *Fame at Last: Who WAS Who According to the New York Times Obituaries*. Kansas City, Mo.: A. McMeel Publishing, 2000.

Baranick, Alana, Stephen Miller, and Jim Sheeler. *Life on the Death Beat: A Handbook for Obituary Writers*. 2nd edition. Oak Park, Ill.: Marion Street Press, Inc., 2005.

Beeson, Trevor, ed. *Priests and Prelates: The Daily Telegraph Clerical Obituaries*. New York: Continuum, 2002.

Bishop, Edward, ed. *Daily Telegraph Book of Airmen Obituaries*. London: Grub Street Press, 2002.

Calhoun, Chris, ed. *52 McGs.: The Best Obituaries from Legendary New York Times Writer Robert McG. Thomas Jr.* New York: Scribner, 2001.

Chalmers, Robert. *Who's Who in Hell*. New York: Grove Press, 2002.

Clark, Roy Peter, and Christopher Scanlan, eds. *Best Newspaper Writing 20th Anniversary Scrapbook*. St. Petersburg, Fla.: The Poynter Institute, 1998.

Collins, Billy. *Nine Horses: Poems.* New York: Random House, 2002.

Conniff, Richard. "Dead Lines." *Smithsonian*, October 2003.

Davies, David Twiston, ed. *Canada from Afar: The Daily Telegraph Book of Canadian Obituaries.* Toronto: Dundurn Press, 1996.

———.*Daily Telegraph Book of Military Obituaries.* London: Grub Street Press, 2003.

Fry, Don, ed. *Best Newspaper Writing 1987.* St. Petersburg, Fla.: The Poynter Institute, 1987.

Glover, Stephen, ed. *Secrets of the Press: Journalists on Journalism.* London: Allen Lane, 1999.

Harvey, Sam. *High Adventure: Porter Harvey and the Advertiser-Gleam.* Montgomery, Ala.: Black Belt Press, 1997.

Hegan, Ken. "Famous Last Words." *Toro*, September 2004.

Hume, Janice. *Obituaries in American Culture.* Jackson: University Press of Mississippi, 2000.

Lende, Heather. *If You Lived Here, I'd Know Your Name: Notes from Small-Town Alaska.* Chapel Hill, N.C.: Algonquin Books, 2005.

Massingberd, Hugh. *Daydream Believer.* London: Macmillan, 2001.

Massingberd, Hugh, ed. *A Celebration of Eccentric Lives.* London: Macmillan, 1995.

———. *Second Book of Obituaries: Heroes and Adventurers.* London: Macmillan, 1996.

———. *Third Book of Obituaries: Entertainers.* London: Macmillan, 1997.

———. *Fourth Book of Obituaries: Rogues.* London: Macmillan, 1998.

———. *Fifth Book of Obituaries: Twentieth-Century Lives*. London: Macmillan, 1999.

———. *The Very Best of The Daily Telegraph Books of Obituaries*. London: Macmillan, 2001.

New York Times staff. *Portraits 9/11/01: The Collected "Portraits of Grief" from the New York Times*. New York: Times Books, Henry Holt and Co., 2002.

Newsday staff and the Tribune Company. *American Lives: The Stories of the Men and Women Lost on September 11*. Philadelphia: Camino Books, 2002.

Powell, Georgia, and Katharine Ramsay. *Chin Up, Girls!: A Book of Women's Obituaries from the Daily Telegraph*. London: John Murray, 2005.

Siegel, Marvin, ed. *The Last Word: The New York Times Book of Obituaries and Farewells, a Celebration of Unusual Lives*. New York: William Morrow and Co., 1997.

Singer, Mark. "The Death Beat." *The New Yorker*, July 8, 2002.

Smith, Martin. *The Daily Telegraph Book of Sports*. London: Pan, 2001.

Talese, Gay. "Mr. Bad News" in *The Gay Talese Reader: Portraits and Encounters*. New York: Walker, 2003.

Whitman, Alden. *Come to Judgment*. New York: Viking, 1980.

Acknowledgments

═══

There is no book here without these obit writers and editors and their wonderful words, which I quote freely in these pages. I admired many more than I could interview, and interviewed more than I could include. I'm grateful to each of them. This would be a different book without the contributions of Jim Nicholson. He was as generous to this writer as he was to the 6,000 or so people whose full-blown obituaries he wrote through the years. I am indebted to all of my sources, particularly those who talked to me repeatedly: Chuck Strum of the *New York Times*, Andrew McKie of the *Daily Telegraph*, Adam Bernstein of the *Washington Post*, Trudi Hahn of the *Star Tribune*, Larken Bradley of the *Point Reyes Light*, and Stephen Miller of the *New York Sun*. Nigel Fountain and Caroline Richmond went out of their way to show me London. Along with the many obituarists who appear in these pages, Sheri Baxter, Liz Renzetti, and Mel Watkins were eloquent guides. Dr. Nigel Starck shared his research; Tim Bullamore, Ralph Cipriano, Joan Harvey, Colin Haskin,

Acknowledgments

Lonnie Hudkins, Andrew Losowsky, and Gayle Ronan Sims contributed much to shape my thoughts about obituaries. Kay Powell of the *Atlanta Journal-Constitution* found me the obit of wine merchant Jim Sanders, mentioned in the last chapter, then kindly tracked down Sanders's biographer, Doc Lawrence, who furnished further details.

The golden age of the obituary glitters partly thanks to its fans, in particular, Carolyn Gilbert and her Association of Obituarists, Thomas C. Hobbs, Dr. Cory Franklin, and Amelia Rosner, Bill Schenley, and their cohorts at Google groups' alt.obituaries. Gilbert and Rosner led me to many of the people listed above, and Rosner spent hours guiding my explorations with intelligence and humor.

Thanks to the photographers represented in this book, as well as Mark Blackburn, Dave Mitchell, and Gwen Taylor; to all the friends and colleagues who sent me obits; and to those who aided and abetted the writing of this book: E. Jean Carroll, Jan Cherubin, Princess Clark, Marcelle Clements, Laura Shaine Cunningham, Roger Director, Mark Golodetz, Nancy Hass, REL, Leonie Maclean, Becky Okrent, Nancy Ramsey, David Robbins, and Harriet Rochlin. Paul Slansky's bus plunge headlines are like everything he collects: priceless insights into the culture. Melinda Marshall helped me to find my voice in the proposal, and Susan Squire not only edited the proposal but spent countless hours with the manuscript, helping me find the story. Terry Bazes, Kate Buford, Benjamin Cheever, Esmeralda Santiago, and Larkin Warren shared their considerable editorial talents during the reading of many versions of these pages, and made the process a pleasure. I've been lucky to work with some of the best editors in New York: Susan Bolotin, Mark Bryant, Judy Daniels, Robert Friedman, Jay Lovinger, Dan Okrent, Pat Towers, Carrie Tuhy, and Terry McDonell, who assigned numerous celebrity tributes to me while at US, but was prevented by fate from publishing any of them. I am grateful to Betsy Carter, who published an early version of "I Walk the Dead Beat" in *New York Woman;* Dave Daley, who published a version of the beginning of "Tributes" and my sendoff of Marlon Brando in the *Journal News;* and Bob Sullivan, who published my tribute to Hepburn in the commemorative *LIFE Katharine Hepburn: 1907–2003.* Amanda Urban launched me as a magazine writer, and I'll always be grateful to her and to Liz Farrell of ICM.

Chris Calhoun was such a dedicated fan of Robert McG. Thomas Jr. that he collected his work in the book *52 McGs.* I've had the good luck to count as my agent someone who is a hero to obit lovers.

David Hirshey believed in this project from the beginning. I'm grateful for his confidence, experience, and wit, and thrilled to be on his team. Nick Trautwein has been an invaluable sounding board and tireless with assistance in matters large and small. Thanks to Ed Cohen and David Koral for their attention to every detail; they and the rest of the HarperCollins staff have been the ideal publishers.

My personal thanks to Dave and Dorothy Johnson; Evelyn and Jules Fleder; Jackson, Carolyn, and Nicholas Fleder; and Rob Fleder, without whom this book would not exist—I am so lucky to have his discerning eye, his sense of where the uncharted work needed to sail next, his delight in the trip, and his steadfast love throughout. . . .

How many people it takes to sustain the work of a book—and I haven't yet mentioned the people whose deaths inspired the obituaries. They endure in the pages of newspapers, in cyberspace, and in this and other books. They and their survivors have our humble gratitude for all they've taught us in life and death.